No. 1727
$17.95

THE GIANT HANDBOOK OF FOOD PRESERVING BASICS

BY ELIZABETH & ROBERT WILLIAMS

TAB BOOKS Inc.
BLUE RIDGE SUMMIT, PA. 17214

Also by the Authors from TAB Books Inc.

No. 1597 *Building with Salvaged Lumber*

All photos are by the authors Williams, III.
and by their son, Robert L.

Mason Jars is a registered trademark of the Ball Corporation.

FIRST EDITION

FIRST PRINTING

Library of Congress Cataloging in Publication Data

Williams, Elizabeth, 1942-
The giant handbook of food preserving basics.
Includes index.
1. Food—Preservation. I. Williams, Robert Leonard, 1932- II. Title.
TX601.W67 1984 641.4 83-24347
ISBN 0-8306-0727-7
ISBN 0-8306-1727-2 (pbk.)

Cover illustration by Al Cozzi.

Contents

Introduction

During the decades of plenty following World War II, there would have been little reason for this book to be written. There was prosperity virtually everywhere in America; most people enjoyed wonderful produce at reasonable prices at supermarkets and roadside stores. The emphasis on ecology and sensible nutrition—plus the ominous messages concerning carcinogenic ingredients in our food supplies—had not yet received national attention, and the alarm bells had not sounded.

Now, with thousands of people insisting on sodium-free diets, low cholesterol cooking, weight-control, and better nutritional values, the pendulum has swung again in the direction of the home food preservation endeavors. The do-it-yourself movement has had far-reaching effects, and millions of citizens are pursuing, through economic necessity, the pleasures of the work, or similar motives, the challenges and rewards of a return-to-the land life-style, even if the return is only to a postage-stamp sized garden at the back of the lot.

To their delight, these gardeners, from every imaginable life-style, are learning what they have missed for so long: the inestimable delights of fruits and vegetables straight from the garden or orchard.

The movement from the rural life-style to the urbanization existence reached its peak a decade or so ago. Now more people are returning to the outskirts of the cities, suburbs, small towns, and weekend or seasonal retreats, and increasing numbers of these people are rediscovering the manifold joys of growing and eating their own food. They are often restricted, however, to the enjoyment of these delicacies only in season. During the off-season they revert to canned goods from stores and wait impatiently until they can again plow, plant, hoe, and wait for the first spring crops to mature.

This waiting is unnecessary, however. It is possible to preserve foods at home in a variety of ways, and these canned, dehydrated, cured, or frozen foods are often nearly as flavorful, nutritional, and visually appealing as the fresh foods. In many cases, the preserved foods are actually better in flavor than the fresh crops.

Home food preservation is easy, convenient, inexpensive, and enjoyable; there are nutritional and aesthetic, as well as general health, values involved.

Years ago we were canning and freezing, like countless others, the traditional foods. Then, as we started to experiment with newer (to us, at least) foods, we also began to experiment with new methods in our preservation endeavors.

We tried the yard-long cucumbers, the blue potatoes, and the winged beans, which were supposed to be the final solution to the gardener's dreams. But, to our delight and profit, we found that the best experiments were in the areas of food preservation.

For instance, we had heard and read in many booklets that it is impossible to dry or dehydrate watermelons, cantaloupes, tomatoes, and numerous other foods that we enjoy so much. But we tested, retested, and revised our approaches and plans. We learned that a watermelon which is barely palatable when freshly cut is an indescribable delicacy when dried; a great fresh melon is beyond comprehension. The same is true of several other fruits and vegetables.

We began trading suggestions—and some secrets—with friends and neighbors. Our mild interest began to grow into a controlling fascination. Soon we found ourselves discussing the best methods of canning or freezing whatever foods our acquaintances—and often strangers whom we happened to engage in conversation—grew or enjoyed from the garden of someone else.

Then we were asked to teach classes for continuing education programs in the community college areas so that others could join our ranks. Unfortunately, there were no books that really suited our needs.

So we decided to write one, and this book is the result of that decision. We are often reminded of Dr. Faustus in Christopher Marlowe's drama: this man, who had the opportunity to have any wish granted or any fantasy fulfilled, quickly chose to have grapes out of season.

We understood, in part at least, how he felt. We wanted virtually everything we grew, in season and out. And this book is, in part, the result of our own desires and needs.

There are, however, many other reasons. Chief among these are, partly in view of recent alarming national incidents, the desire to have food that we know to be safe, free from contamination, and picked at the peak of their growth cycle after having been grown under controlled conditions—controlled primarily by us and by Mother Nature—as well as food that is economical, handy, and delicious.

There is considerable pleasure and security in the knowledge that in the freezer, on the cabinet shelves, and in jars in the pantry are our grocery supplies for the next several months, even years. It is reassuring to know that, if there are food shortages— which, by the way, we do not predict—or economic structures— which none of us can be certain of controlling—there is at least food.

In the preparation of this book we have had help from so many individuals that to name any would be to emphasize the omissions of names of persons equally helpful. So in one wide-spread gesture we wish to thank our friends, our neighbors, and our fellow gardening enthusiasts everywhere. At the same time we would like to welcome into the fraternity those newcomers (our young son Robert III is among them) who are just beginning to realize the joys and physical benefits of growing and preserving foods of all types.

It may have all started, centuries ago, with the desire to have grapes out of season, but today it is possible to have nearly every imaginable food in and out of season, and there is no need to barter one's soul for the privilege. All you need is a source of food, whether it be your own garden or the generosity of others, and either love or need, preferably both.

In his marvelous poem "Two Tramps in Mud Time," Robert Frost passed on his observation that "only when love and need are one and the work is play for mortal stakes is the deed ever truly done for Heaven's and the future's sakes."

So we are further indebted to Frost, who also

told us of the joys of remembering something you never knew you knew.

We always knew how much we enjoyed our own efforts and results in food preservation, and we enjoyed even as much, perhaps more, the pleasures of exchanging these data and helpful hints with others. We also found, to our delight, that it can be stimulating and rewarding to pass these bits of knowledge and opinions on to others through this book.

Our foremost wish is that those of you who read and benefit from this book will enjoy continued success and happiness in growing, preserving, and eating the best foods that Nature and know-how can produce.

1

Principles of Food Preservation

All drying, canning, curing, and freezing processes are based on very simple principles: enzyme control, destruction of bacteria, utilization of bacteria, and oxidation. All of these principles can be applied easily for nearly all types of foods as long as basic precautions are observed.

Perhaps the most universal problem is that of enzyme control. An enzyme is a protein-like substance formed in cells of plants and animals. These enzymes act as chemical catalysts that, in turn, cause a chemical reaction to accelerate or become unstable. When enzyme-caused reactions occur, fruits may become soft, grains and potatoes will sprout, and meats and dairy products will liquefy.

Enzymes are absolutely essential for fruits and vegetables to ripen. The very chemical processes which allow ripening also cause spoilage, however, since the ripening processes continue even though the fruit has reached its peak in flavor and appearance. Blanching the food by steaming it or boiling it for only a few minutes will effectively halt enzyme action.

Enzyme action seems to accelerate when fruit or vegetables are picked and also when they are sliced or broken. For this reason it is important to process the food as quickly as possible—one argument for having your own garden so you can be certain just how old the food is. If you buy your produce at the supermarket or the roadside stand, even though the food may look good, it is also possible that it has been picked for several days, during which time invisible enzyme action has advanced to the point that the food has reached the precarious stage.

The other type of food spoilage is the result of the work of micro-organisms: yeast, mold, and bacteria. Here the problem is more significant, but it can be easily handled. You can simply prevent the growth of the micro-organism, or you can kill it. To grow, the micro-organism needs three essential elements: oxygen, moisture, and heat. If you take away any of the three, you in essence solve the difficulty.

Drying will not kill bacteria; it simply inhibits growth by taking away the needed moisture. As soon as the dried food is returned to open air, the moisture in the air again provides the needed water and the bacteria begin to work anew. For this reason dried foods should be kept sealed in air-tight containers until it is time to use them.

If you freeze foods, you take away the heat needed for bacterial growth. Again, though, when the food is allowed to thaw, the growth starts again, and soon the food is ruined.

If you can your food, you actually destroy the bacteria. You must, however, take every precaution to see that the food is not reinvaded by bacteria, and you can prevent the problem by making the container air-tight.

Yeast cells feed on sugar and other carbohydrates in foods, and, while many times yeast cells perform a useful function, the result is often the production of undesirable odors and flavors in many foods. Yeast cells stop growing when food temperatures drop near the freezing mark. Similarly, yeast cells can be destroyed usually at temperatures of 180 degrees.

Molds are in many senses similar to yeast cells. These growths are recognized by the fuzzy, dark-colored patches that appear on breads and other types of foods. Molds can be destroyed easily by raising the temperature of the food to 180 degrees.

The final type of growth that causes trouble in foods is bacteria. While yeasts and molds can cause food to have a foul taste, some bacteria can cause severe illnesses or even death. The bacterial cells are also by far the hardest to destroy in foods. Unlike yeasts and molds, bacteria can cause food spoilage even though there is no discernible odor or taste. Consequently many people ingest the damaged food and become severely ill before they realize that the food was contaminated. Some bacteria pores can be killed only by processing food at temperatures of 240 degrees for as long as 30 minutes to 1 hour.

Under certain conditions, bacterial growth is desired and necessary. The most obvious area is in the process of soil decomposition, where bacterial action is largely responsible for the formation of topsoil rich in the nutrients needed for good plant growth. In the food preparation areas, certain bacterial actions occur in the fermentation processes. These growths are part of the conversion of sugar to alcohol, the formation of cheeses, and the preparation of breads and cakes.

A mild problem facing the person preparing fruits and certain vegetables for canning or freezing is the action called *oxidation*, in which fruit rapidly discolors to produce an undesirable brownish hue. While no real harm is done, there is a definite loss of aesthetic appeal. To prevent oxidation, you can drop apple or peach slices into water or into an ascorbic acid mixture. Of course, speed in preparation is one of the best safeguards against oxidation.

Two basic principles of home food preservation must be remembered. You must halt enzyme growth in order to prolong the life of the food to be processed, and you must either kill the microorganisms present in food or inhibit their growth. You can accomplish these goals in several ways, all of which will be discussed in Chapter 2.

2

Basic Techniques of Food Preservation

Canning, like any other form of home food preservation, is subject to many variables. You cannot use fruit or vegetables that are overripe, inferior in taste, and grown too rapidly for formation of the flesh, and then expect your canned goods to taste like delicacies. Canning does not improve food, and, with very few exceptions, neither can freezing, drying, or any other method of preserving.

The economics of home food preservation must also be questioned, especially if you plan to drive to the grocery store and pay market price for your food, then pay the costs of canning or freezing, and then add to the expense your time and energy.

Home food preservation must also be seen in the light of your geographic area. If you live in a climate that permits a year-round garden, you may want to give canning and freezing second thoughts.

If you are like most Americans, however, you must either grow and preserve your winter's food between June and October, and the rest of the year eat what you have 'put up,' or you must buy your food.

You actually do the vast majority of your preserving in about six weeks. From late August or September, in many instances, the garden plot is fallow until the following March or April, when it is plowed and seeded. It is often as late as June before the first crops are ready for harvest, and, unless you have a long enough summer for two gardens, you may be finished with your preserving by the end of July. Even if you have two gardens and harvest up until frost, perhaps afterwards, you still have a long period without fresh food from your garden.

In terms of the dollars and sense of gardening, you must include the cost of seeds and fertilizers, if any, the cost of driving to the store for your seeds, the cost of gasoline to run the tractor or tiller, the costs of freezing bags, electrical power to run the freezer, jars, lids, and rings, and the cost of operating the stove or range to process the food.

Obviously, when balanced against the joys of the gardening, the pleasure of watching the growth, the delights of picking, and the even greater pleasure of eating or simply sitting and admiring the neat rows of beautifully processed containers of food, the economical angles fade quickly. Be wary of the false economy of home food preservation, but be equally, if not more so, delighted by the great money-saving activities that you can enjoy if you garden intelligently.

When the garden crop is gathered and it's time to preserve the fruits or vegetables, you must decide how you will preserve your food. There are five very simple methods of preserving food and four slightly more complicated ones, although none of the methods is really difficult.

The simple methods are canning or bottling, moisture reduction, freezing, fermentation, and simple storage. The slightly more difficult methods are curing, smoking, sulfuring, and pasteurizing.

CANNING

Canning, or ***bottling***, is one of the oldest and most reliable methods of preserving food. In many areas ***canning*** meant the actual processing of food in metal cans, but generally the term applies to the storing of food and juices in air-tight bottles, Mason canning jars, or other types of glass containers (Fig. 2-1).

Your objectives are to heat the food in the glass jars until you have inactivated the enzymes and killed the bacteria, and then to seal the jars hermetically so no other bacteria can contaminate the food. Seal the jars before you process the food or immediately afterward while the food is still extremely hot. As the jar and its contents cool, there will be a lower air pressure, and the vacuum principle will cause the lid to be sucked suddenly against the opening of the jar. The vacuum tendency inside the jar will hold the lid tightly in place until you are ready to use the food. If this vacuum sealing doesn't occur, or if somehow the vacuum is broken, the food will spoil quickly.

It is a good idea to check jars and lids carefully

Fig. 2-1. A shelf filled with canned goods is one of the rewards for diligent work in canning. The well-filled jars shown here will cut several dollars from the grocery budget.

before you store them to be certain that the sealing occurred properly. You can hear the sealing take place virtually all over the house.

If you don't hear the sealing, you can usually tell simply by looking whether a jar has sealed properly. Hold the jar up to eye level, and you can see the slight indentation or sunken area in the center of the lid.

If you neither heard the sealing nor were able to determine by visual inspection that a jar sealed, you can simply push down on the lid rather firmly. If the lid gives under the pressure and then clicks back upward when you remove your finger, the jar did not seal. It should be processed again quickly or stored under refrigerated conditions until you can either reprocess or serve the food soon.

Occasionally a jar will apparently seal, and then later the lid will pop back up. If this should occur within a few hours of the processing, you can still use the food. If you notice the problem a day or more later, however, discard the food.

The most important aspects of canning are related to the temperature of the food during processing and the acidity level of the food. If the foods are high-acid, you can use a basic water-bath or open-kettle method of canning. If they are neutral or low-acid, you should use a pressure canner.

In high-acid foods the micro organisms are, for some reason, more easily killed, and the acid content of the food prevents bacteria from developing afterwards; so you can process the food at the basic boiling point, but you must maintain the heat for a prolonged period of time. You must also consider how the food was packed. If you have large sections of the fruit or vegetable, such as peach halves, the heat tends to reach the center of the jar very quickly. If, however, the food is pureed or packed into a near-solid mass, such as cooked pumpkin, the heat takes longer to reach the center, and if you don't process for several minutes, there will not be enough heat at the center of the jar to kill all of the micro-organisms.

The water-bath processing and open-kettle methods are not recommended for low-acid or neutral foods, such as green beans, squash, and potatoes. The water-bath method does work well for high-acid foods like tomatoes, apples, and most other fruits.

Water-Bath Method

In the water-bath method, you fill your jars, all of which have been washed in hot water before using, with the food up to 1/2 inch below the top of the jar. Then fill the jar with boiling water—or liquid from the cooking of the food—until it covers the food. Screw on rings after the lids have been placed on the jars. Place the sealed jars in a container filled with water. Place over heat and bring the water to a boiling point for low acid foods and to at least 150 degrees for high-acid juices.

See Chapters 7 and 8 for times and temperatures for all types of fruits and vegetables.

The low temperatures for fruit juices are sufficient to prevent food from spoiling. Juices, if heated to the boiling point, tend to lose a great deal of their flavor.

Keep in mind when you are canning low-acid foods that botulism, one of the deadliest poisons known to man, can contaminate foods unless the processing temperatures are 240 degrees for a minimum of 15 minutes.

Open-Kettle Method

In the open-kettle method of canning—also not recommended for low-acid or neutral foods your success is dependent on the high heat of the food, the jars, and the lids used. First place clean containers in a large vessel and completely cover the jars with water. Bring to a boil and continue boiling for 20 minutes. Keep the jars hot while preparing the food to be canned. Lids and rings to be used should also be placed in boiling water until you are ready to seal the jars with them. The food must be completely cooked and kept boiling during the the entire canning procedure. Jars should be filled to the bottom of the neck, wiped with a clean cloth, and closed with the hot lids and rings. As the contents cool, the vacuum effect is created, and the lids are sucked downward to create an hermetic seal.

In the open-kettle method, as with all types of

canning, you should leave canned food out so that you can observe it for a day or two. If lid pops back up, you know there was not a proper seal, and the jar and its contents should be used at once or discarded.

Pressure-Canning Method

To can low-acid foods, such as green beans, use the pressure-canner method. A pressure canner is a large pan or vessel which can be tightly sealed by means of a fitting lid and a rubber gasket. When lid and gasket are in place, the only way for steam to escape is by way of the valve located in the lid. The theory of the pressure canner, like that of the pressure cooker, is to trap steam under pressure.

You process the food in terms of pounds of steam pressure rather than by air pressure, and you can easily achieve the heat equivalency of the necessary 240 degrees needed to kill all micro-organisms. A dial on the top of the lid indicates the amount of pressure generated. You need to watch the dial fairly closely to see that the steam pressure does not fall below the recommended point during the processing. If you should allow the pressure to drop more than a pound or so for a minute or two, we recommend that you start the timing over and cook or pressure longer. The added time in the canner will not damage the food if it is for only a few minutes, and it certainly will make the food safe beyond doubt.

If the pressure exceeds the recommended amount, you can simply reduce the heat until the pressure drops. No harm is done if the pressure goes up a pound or two, and the vent or petcock on the lid allows steam to escape so that there will be no real danger from the extra pressure. If the pressure gets too high, you can always lift the canner off the heat until the pressure returns to normal.

MOISTURE REDUCTION

In the moisture-reduction methods of food preservation you will be doing one of two things: drying or dehydrating. There is a technical difference in the two terms, although many people use them synonymously.

If you ***dry*** foods—by using the solar drying method—you simply spread the food, after slicing it into thin slivers, in the open air where it can get a great deal of sunshine and air. You spread cheesecloth or put screen wire over the food to protect it from insects and birds. When it reaches the proper state of dryness, you bring the food inside and store it in air-tight containers (Fig. 2-2).

In the ***dehydration method***, you extract the

Fig. 2-2. You can shell out the dried sunflower seeds by holding the seed pod firmly in one hand and flipping the seeds with the thumb of the other hand. Use a large towel or cloth to catch the seeds, since they tend to fly in several directions.

moisture from the food by artificial means. There are several types of food dehydrators on the market, and you can make one for a very low cost, if you prefer.

In the dehydrator there is a source of heat—usually an electric coil, a light bulb, or heat from a conventional gas or electric range oven. There should be a way in which air can be circulated by a convection method. When the food is properly dehydrated, you remove it from the oven or dehydrator and pack it in air-tight containers as you did with the sun-dried food.

There are some obvious advantages and disadvantages to either method, but there are very minor problems with any forms of dehydration or drying. With solar drying, you obviously need a sunny day with temperatures, for best results, in the nineties. If you have your food prepared and rain begins, you are thwarted, at least temporarily. If the rainy session lasts for several days, the food may be ruined.

Another disadvantage is that if the weather is unsettled, you may have to stay close to home during the drying process; so you will not be able to work, shop, or visit for prolonged periods. Solar drying also takes longer, often several days, for complete moisture reduction.

One of the major advantages of solar drying is that you use no power or other fuels, thereby saving on the actual cost of processing. You do not have the expense of buying a dehydrator, and you do not have the oven tied up for most of a day.

The commercial dehydrators have several advantages. First of all, they are relatively inexpensive—from about $20 to $150. They use very small amounts of power—some cost only about $.05 a day to operate. They are also highly efficient, clean, and quiet.

The major disadvantages of dehydrators, in addition to the fact that you must buy them, are that they are usually fairly small, and only a reasonably small amount of food can be processed at a time. Many models have a drying space of 4 square feet. Since there must be air space between racks, you can dry far less food that you can in the solar method.

The kitchen range oven works well, but you cannot use the oven for baking during that time. Usually you must also have a small fan located near the oven and the oven door cracked slightly, so that the air in the oven will be properly circulated.

Considered from several viewpoints, however, any form of dehydrating is a superb form of food preservation, and the results are delightful. Among the many advantages of dehydrating is the conservation of shelf storage space, since all dehydrated foods shrink considerably—some up to 1/15 or even less the original size. You also do not have to worry about a vacuum seal, but the container must be air-tight since room moisture can cause the foods to begin to rehydrate and spoil. Some foods actually taste much better after dehydration than they did in their original raw state. In many instances you do not have to rehydrate the foods before you eat them, and you do not have to cook in order to assure safety from micro-organisms, since these bacteria must have moisture in order to live.

It is also possible to dehydrate or dry several foods that really cannot be preserved effectively in other methods. Watermelon and cantaloupe can be dried easily, and the taste is incredibly delicious. There are few, if any, other methods by which these two melons can be preserved. Surprisingly, melons are among the foods that actually have a better flavor after preserving than they had while fresh. Even a poor watermelon, as long as it is in sound condition, becomes delectable after dehydrating (Fig. 2-2).

FREEZING

The principle behind the freezing of foods is that micro-organisms cease to grow when temperatures fall below zero degrees Fahrenheit. The frozen state of the food prevents further ripening or decomposition, and the bacterial growth is completely retarded.

The major advantages are that frozen foods retain most of their original flavor and nutrition, little preparation is necessary before storing, most foods can be frozen effectively, and entire meals or wholly cooked dishes, such as casseroles, desserts,

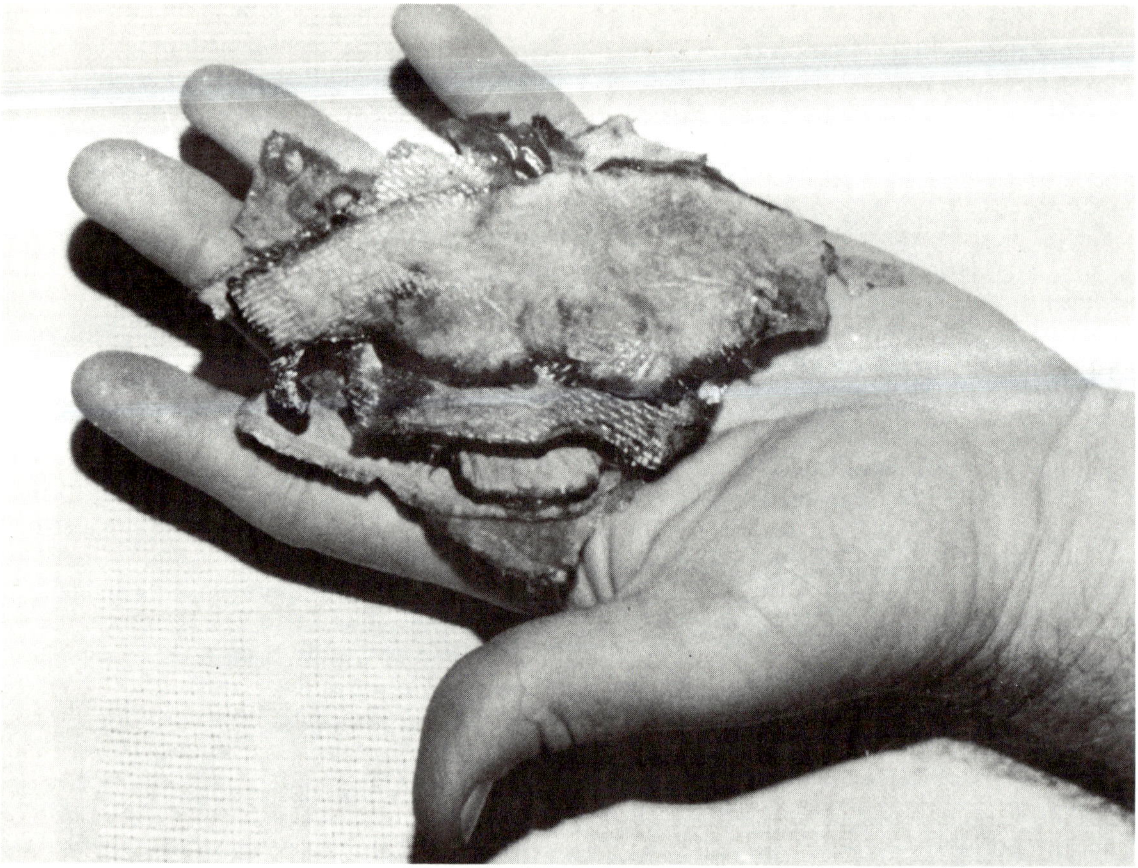

Fig. 2-3. Dried watermelon is only a tiny fraction of the original size of the melon. Here, a huge slice of melon has been reduced to a small handful, but the dried melon flesh is delectable.

and home versions of TV dinners, can be preserved.

The major disadvantages are the cost of the freezer, the danger of thawing and spoiling during a prolonged power failure or mechanical breakdown, and the cost of operating the freezer 24 hours a day.

Freezing is one of the most universally used forms of home food preservation, and virtually dozens of types of foods can be frozen easily. Among those that cannot be frozen well are those foods high in starches, such as potatoes, and those valued primarily for their crispy texture when eaten raw, such as celery and lettuce. Foods so high in moisture content that the water freezes, and causes the fibers to rupture, with the result that the food collapses upon thawing also do not freeze well. Melons, tomatoes, and raw cucumbers are some examples.

Without a doubt, freezing is the best way to preserve many foods. Meat in particular can be frozen easily and kept for months without damage or loss in food value or taste. Many fruits are marvelous to freeze, and many people like frozen vegetables better than canned or dried ones.

FERMENTATION

Fermentation is a process through which organic substances are transformed into simpler compounds as micro-organisms cause a chemical breakdown. In some processes, sugar is transformed into alcohol; in others, starch is transformed into alcohol; in others, starch is split into sugar meltose. Carbon dioxide is produced in many processes in such quantities that the bubbles rising to the surface give the suggestion of boiling water. In fact, the word fermentation comes from a word meaning "boiling."

While beer, ale, wine, and other alcoholic beverages are products of a fermentation process, many other foods which are non-alcoholic also ferment. The yeast process used in making wine is also used to make dough rise; so breads are products of fermentation. Many pickles and such stand-by foods as sauerkraut are fermented. Beans, eggs, and beets, as well as many kinds of fruits and vegetables, are also fermented, in a sense, during the pickling stages.

When making wine, you need either wild or commercially prepared yeast, either natural sugar present in the fruit or granulated sugar, and fruit pulp and juices. If you pour the juice, sugar, and yeast mixture into a large container and place a vented stopper, so that carbon dioxide can escape but air cannot be introduced into the mixture, you actually have a "living" organism, such as Pasteur described as "life without air." The enzyme action starts the process of conversion of sugar into alcohol. When the "working" has been completed and the aging time has elapsed, you have wine. Chapter 16 gives instructions on wine-making.

When you ferment or pickle foods, you depend usually on a salt or brine solution. You place the food to be picked into a container and pour the brine over the food, then close the container and allow the pickling process to occur. Chapter 8 tells how to make various pickles.

ROOT CELLAR STORAGE

Certainly root cellar storage is among the simplest forms of home food preservation. If you have a basement area that can be used, you can store large amounts of food for several months, and you have to make very few preparations. The best foods to store in root cellars or basements are Irish potatoes, sweet potatoes, onions, cabbages, carrots, and many other root crops.

For short-term storage you can keep apples, winter squash, vegetable spaghetti, and several other types of melons. You can store nuts, insulated green tomatoes, and numerous fresh garden vegetables in the cellar or basement as long as the food is kept away from dirt or concrete and as long as there are no mice or insects to attack the food.

Nuts and peanuts—which are tubers rather than nuts—can be kept indefinitely in a cellar or basement as long as they don't absorb moisture to an excessive extent.

The primary advantages of basement or root cellar storage are that no power is used, no extensive preparation is necessary, and the temperature and light conditions are ideal.

One disadvantage is that insects and enzyme action can cause food to decay, and one rotten apple will, in fact, spoil the barrel. The decay will spread quickly, and the resulting odor lingers for weeks. The shelf on which the damage occurred may well be so badly contaminated that a severe scrubbing and disinfecting may be necessary.

The few preparations for cellar storage are simple. You may want to spread a sheet of plastic on the floor in order to keep moisture from the food, and you will want bins, shelves, or racks on which or in which to store the food. You may also want to sprinkle lime over potatoes in order to discourage insects and enzyme action.

CURING

Meats can, with few exceptions, be frozen, canned, or dried quite easily. Some people, however, prefer to cure their meat, particularly pork, since curing tends to preserve well and to add a flavor to the meat that is not present under any other conditions. Cured ham, for instance, pleases most people better than fresh meat.

Briefly stated, ***curing*** is the application of a spoilage preventative to the meat to keep bacterial action from occurring. The two major ingredients in the curing process are salt and sugar.

The salt destroys bacteria, but too much salt can be harmful to the meat itself. Hams that are cured solely by salt can become so hard that it is very difficult to cut the uncooked meat even with the sharpest knife. The cooked meat is often too salty to be really palatable, and the hardness of the meat remains an undesirable condition.

If sugar is added to the salt, the result is a meat that is protected against bacteria and also highly tasty and generally tender. When you read or hear of sugar-cured hams, though, you should realize that the processors used both salt and sugar.

You can cure ham and other pork by rubbing salt by hand into it. You must, however, keep the meat at a low temperature while the meat is curing. You can also use a needle to inject the brine into the ham, but the hand-rubbed ham is most popular.

After the salt and sugar are thoroughly rubbed into the ham, the ham should be wrapped in brown paper and then enclosed in a sack and hung up to cure in a room that is cool and dry.

Other types of curing are very similar, and these will be discussed in Chapter 13.

SMOKING PROCESSES

Many meats, particularly pork products, are smoked after other types of curing have been completed. The basic method is to have a small building called a smoke house in which the meat may be hung or stored on shelves. Inside the building a fire fed by green hickory chips or blocks burns constantly.

Several kinds of fish and poultry, and occasionally beef, can also be smoked if you desire the smoky flavor. To prepare the meat, you may hand-rub it with an equal mixture of salt and sugar, or you may brine it by soaking the meat in a solution of salt and water plus small amounts of brown sugar and saltpeter. A good workable mixture is 10 pounds of salt, 4 pounds of sugar, 4 or 5 tablespoons of saltpeter, and 10 gallons of water.

After you have soaked the meat—at the rate of four days for each pound of meat—you can hang it in the smoke house. If you want a faster method, you can inject the brine solution into the meat with a special needle. If you inject and soak, you can soak the meat for about 8 hours and the results will be satisfactory.

The smoke house can be any sort of small building that has a trench under it for the fire and room inside to hang the meat. This method will be discussed more fully in Chapter 13.

SULFURING

Sulfuring is a method of treating fruits before or during drying. To sulfur, you buy commercially prepared sulfur and burn it and let the smoke penetrate the fruit to be dried. The sulfur acts as a mild preservative and affects the taste of the fruit slightly.

We have found that the use of sulfur is only an inconvenience, and we see no real reason to use the process, in spite of claims that sulfuring helps fruit to retain its color, repels insects, prevents mold, and discourages bacterial action.

We have eaten fruit dried through the sulfuring method and fruit dried with no treatment, not even blanching, and we personally enjoy the unsulfured taste better.

If you wish to try the sulfuring, however, purchase elemental sulfur that is a 95 percent purity mixture. Stack trays of fruit so that smoke and air can pass between them. Burn the sulfur, always outside—you don't want to breathe the fumes for very long—in a metal container, such as aluminum pie pan or empty coffee can. Let the smoke filter around the fruit. Use about 1 tablespoon of sulfur to every pound of fruit. When you are ready, light the sulfur and place a large cardboard box or other covering over the fruit and burning sulfur and leave it until juice starts to appear in the pit cavity of the fruit or until the fruit is bright-looking or abnormally shiny.

PASTEURIZING

In a few foods you may want to pasteurize by heating the juice or other type of food and then maintaining a constant heat until all fermentation-causing activity is halted or prevented. Milk, for instance, should be heated to 143 degrees Fahrenheit and held at that temperature for 30 to 45 minutes.

Wine and beer can be heated to 135 degrees Fahrenheit and held at that level for 30 minutes. If you boil milk, you will create a curdling effect and thereby ruin the milk itself, and, if you boil wine or beer, the heat will destroy or highly minimize the alcoholic content and ruin the drink.

For the most part, pasteurization is

unnecessary. Dairy farmers and others involved in the processing of milk commercially already have all the knowledge and equipment necessary to handle the job. If you milk your own cow and wish to save the milk for a period of several days or drink it "straight from the cow," you may wish to pasteurize the liquid.

Beer and wine sometimes have a tendency to oxidize or ferment a second time, but the problem is not significant. Except for special needs—some people like to pasteurize their apple cider and other fruit ciders—you don't really need to pasteurize.

3

Necessary Equipment

It is not terribly expensive to set yourself up in the fields of canning, drying, and freezing so that you can easily store a year's food supply, or even more, but it does cost ***something***. The purchase of a freezer is the greatest expenditure. The prices of pressure canners and dehydrators lag far behind.

Costs are subject to rapid and often discouraging fluctuations. The prices of large items, such as freezers, may vary by more than $100 between brands or with the amount of extra conveniences.

CANNING EQUIPMENT

It is a difficult decision when you try to reduce your options to one purchase. Some foods are better canned; others are better frozen; some are more tasty when dried. Even if you opt for canning as your major method of home food preservation, you still have several decisions to make.

In the area of canning, if budget permits, buy the necessary items to can by water bath, open kettle, and pressure, because you can make one judicious decision that will allow you many options later.

Water-Bath Canner

If you buy a water-bath canner, you will also need to buy a supply of jars, lids, and rings. If you buy a pressure canner, you still need the jars, lids, and rings, which are tempered to withstand the boiling temperatures of the water bath and the steam pressure and heat of the pressure canner.

The 21 quart size water-bath canner sells for $12.00 to $25.00 for the steel processors. The less expensive ones are usually made of aluminum.

A word of caution is in order. Do not be misled by the size of the canner. One that is advertised as a 21-quart canner will really hold only 7 one-quart jars; the 33-quart size holds 9 one-quart jars or perhaps 18 one-pint containers. The rest of the canner or processor is water and air space.

If you are going to be involved in home food preservation in a very big way, you can buy a blancher, 8- or 9-quart size, for about $30.00;

however, we have found that ordinary kitchen pots and pans work beautifully.

You will need a rack of some sort with which to lift processed jars of food from the processor, but these inserts often come with the canner-processor.

A case of wide-mouthed pint canning jars will cost you about $4.75, and a case of similar quart jars will cost about $5.50. The rings for such jars, which fit both pint and quart jars, cost about $1.50 per dozen. The price includes the rings and lids.

You can use the lids only one time. Please, do not try to save a few pennies and use the same lids over; you will likely run into improperly sealed jars, spoiled food, and possibly food poisoning. You can, however, use the rings an almost unlimited number of times. As soon as the lids have been seated and sealed properly, you can remove the rings and use them later.

You can also buy lids separately, and the cost is usually about 80¢ for 12 wide-mouthed lids and 60¢ for 12 regular jar lids.

Emptied mayonnaise jars and other supermarket food containers of the sort will work in a water-bath canner, but do not try to use them in a pressure canner. Use genuine Mason canning jars, which have the word MASON imprinted on the jar.

Pressure Canner

If you choose to buy a pressure canner—not to be confused with a pressure cooker—you can buy in sizes from 8-quart to 21-quart sizes. An 8-quart size will can 7 one-pint jars or 4 one-quart jars, and a 22-quart size will hold 7 one-quart jars or 20 one-pint jars. If you can 20 pints of food, you will have to stack or double-deck the jars.

The cost of an 8-quart pressure canner is about $60.00. The 21-quart size sells for about $90.00.

We strongly recommend a pressure canner with a wire basket that fits inside. With this basket, you can place all seven quarts into the canner at one time, and, more importantly, you can lift them all out at one time. If you have only the rack at the bottom, with no handles, you will need to use tongs or pot holders and lift the jars out one at a time. You run the risk of steam burns and of breaking a precious jar of processed food. The wire basket does not cost more, so it is wise to purchase a canner with this very valuable convenience.

If you buy the pressure canner, you can always use the canner, minus the sealing lid, as a water-bath processor, so there is no reason to buy both pressure canner and water-bath canner.

At first the cost of the pressure canner may seem high enough to be discouraging, but if you deduct the cost of the water-bath canner you can see that the final cost is much lower. You can deduct about $20 from the final cost if you buy only one canner.

We recommend the largest size that is feasible for you. You may have a very small family or small garden and conclude that the 8-quart size is adequate. Remember that pressure canning itself requires about an hour for a full run of green beans. It takes about the same time to run seven quarts as it does to run four.

Remember, too, that you can run a larger pressure canner with only one or two jars, if you need to, but you cannot add to the capacity of the smaller sizes. If you have a good supply of beans and you need to make three runs, your total canner time will be slightly over three hours—not including time needed to pick, wash, or string and break beans. At the end of the three runs, you will have 12 quarts from a smaller canner and 21 from a larger one. As you can see, you nearly double your food amount for the same span.

How many jars you should buy and the size depends on your supply of food, the size and appetites of your family, and other variables. If, however, you have a chance to pick up several cases of jars at a very good price, buy them, since you can use them over and they keep for decades if you care for them properly.

DEHYDRATING EQUIPMENT

In order to dry or dehydrate food, you can spend nothing or you can spend $100 or more. If you plan to dry by solar heat, you can build a small dryer for only pennies, particularly if you have access to some scrap wood and a few tools (Figs. 3-1 and 3-2).

Fig. 3-1. A homemade solar dryer can be made by building a box frame and covering the sides with screen wire to protect the food from insects and birds.

Fig. 3-2. You can staple or tack a clean cloth over the rack and spread sliced vegetables on the cloth to dry.

As we pointed out earlier, however, solar drying leaves much to be desired.

If you elect to buy a dehydrator—and we recommend one highly—you should make your decision on type and size by the specific needs of your family.

The cheapest dehydrators cost about $20, and these units do a satisfactory job. There are some, of course, which cost more than $200, but a large part of the price is due to trimmings and advertising rather than to actual performance and value.

A fixed-size dehydrator that holds 4 12-by-12-inch drawers, sells for around $80.00. You can get an expandable dryer that holds up to 12 drawers, each 12-by-12 inches, for about $140. The unit comes with 4 drawers, and each new drawer you add will cost about $12.

It is doubtful whether the fancier dehydrators work significantly better than the basic models, so we suggest one of the less-expensive models, unless you really need the added drawer space.

With a typical low-priced dehydrator, you can fill the shelves or drawers with sliced food, turn on the dehydrator, and spend the rest of the day involved with other matters. You seldom need to check the food until it has been in the dryer for about ten hours. Even then you may just need to move the bottom drawer to the top, while gradually working all other drawers down toward the bottom (Fig. 3-3).

While the smaller units will not hold as much as the more expensive ones, you can dehydrate your food over a period of several days. Every morning you can load the dehydrator with sliced apples, then remove the dried slices that night. You can also load every morning and night, and dry two loads every 24 hours.

A dehydrator is not absolutely essential to your

Fig. 3-3. A modest-sized dehydrator will sit on top of a table or stove and can be loaded easily with sliced or grated foods.

home food preservation needs, but you will be amazed, if you haven't done any dehydrating yet, just how many foods can be dried, and how delicious the finished product can be.

Space is, of course, one of the major considerations. The smaller units occupy only a tiny corner on a shelf, but the larger ones may take up a great deal of counter space.

For oven drying, you need only your oven, a small fan, and a few containers. These containers may be mesh or rod racks, or they may even be aluminum pie pans with holes punched in the bottoms. They can actually be made of clean and dry wood strips, since the oven heat should never be great enough to cause a fire. We don't recommend the wood slats under any conditions, however, since there is always the possibility that someone else in the family may change the oven controls setting.

For air drying you need only a box covered with screen wire and some racks. Here the wooden racks work well, and you can also use metal cookie sheets or any other material that will allow air to circulate freely and will not harm the food.

For storage of your dehydrated food you need only jars and lids or plastic bags. You must make sure that the containers are air-tight.

Remember that dehydrated foods shrink to only 1/4 of their normal size—the actual amount varies considerably with the food itself—so you will need less storage space than you think. Take care to have a dryer rack with grids close enough to hold the slices of food after they have lost their moisture. A slice of pumpkin the size of a silver dollar may be no larger than a nickel after drying, and the smaller pieces may fall through the slats if you are not careful.

FREEZING EQUIPMENT

A home freezer is the greatest expenditure connected with your home food processing. The cost of a freezer will depend, of course, on size, style, make, model, and number of frills. But the lowest priced freezers will cost from $475 to over $600.

Defrosting a freezer can be very hard work, but the extra power needed to run a frostless can be expensive. You choose whether you want to play more or work more.

SECOND-HAND FOOD PROCESSING EQUIPMENT

While you must be very careful about buying used equipment, there are some excellent buys that you encounter occasionally. For example, we saw a used freezer that was on sale for $15. The body seemed to be in good condition, and we asked the salesman what was wrong with the freezer. He told us that the unit had been struck by lightning. We asked an electrician friend to check out the freezer, and he told us that only two wires had been burned in half. We paid the $15 and replaced the wires, and the freezer has served us beautifully for the past eight years.

We also purchased a kitchen range in superb condition for $8.00, and we have used it for eight years also. It is easily possible to pick up excellent bargains if you are willing to exert a little patience, judgment, and caution.

It is at times injudicious, of course, to buy a used item that has truly been abused, but generally you can tell if equipment has been mistreated or used wisely. If you talk with salesmen in appliance stores and let them know that you are interested in buying a trade-in model, you can save a great deal of money.

At flea markets used water-bath canners, pressure cookers, pressure canners, cases of jars, and other items needed in home food processing are often for sale for only a few dollars, at times even under a dollar. Yard sales have notoriously low prices on articles of these types. It definitely pays to remain alert and take advantage of the excellent opportunities and bargains that can be found among the unwanted articles on display.

4

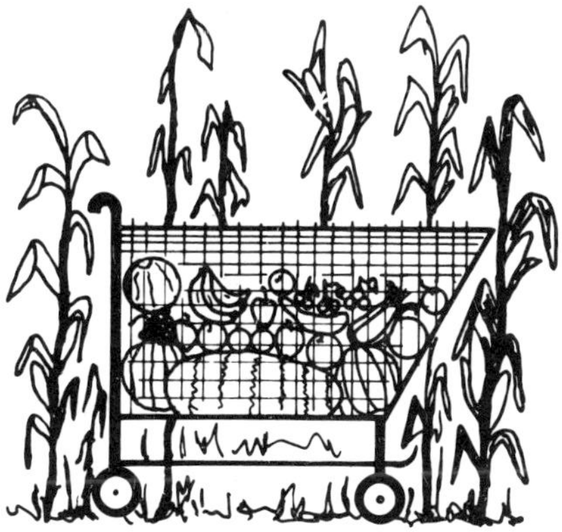

Buying Versus Growing Food to Preserve

The decision to grow or buy your food to preserve depends greatly on many factors. How much garden space do you have? What foods do your family eat most often? How much does food cost in your geographic area? Will you be buying from a supermarket, a local farmer's market, or from private individuals?

How much spare time do you have? Will you need to invest in expensive gardening equipment, or are you willing and able to garden with a shovel, rake, and hoe? Will your health and working hours permit you to spend the necessary time in the garden?

What sort of condition is your garden space in? Will you have to pay someone to plow it for you? Does it need to be cleared? Will you have to pay for all of your seeds and fertilizers?

Will gardening be a one-time experiment or do you plan to make it a part of your life style?

Since each case is different, a few generalized comments may be helpful. Since there are obvious advantages and disadvantages to growing and buying, some effort will be made to point out the major arguments on each side.

BUYING YOUR FRUITS AND VEGETABLES

Several years ago local farmers would load their week's harvest onto a truck and drive up and down city streets selling corn, beans, melons, fruit, and nearly every type of other vegetable that you could want. The growth of supermarket chains, however, soon made food at very low prices available to so many housewives that they found it easier to buy weekly rather than try to can or freeze.

Now, with the high prices of food in supermarkets and inflation in all areas of food production, the small farmer is making a remarkable comeback, and again low-priced food is available in many parts of the country.

Advantages

If you can find ready-to-preserve food in large quantities and encouraging prices at the local farmer's market, you may want to buy rather than grow. There are several advantages to doing so.

If time is extremely important to you, buying food might be right for you. It takes long hours to plow or spade a garden, plant seeds, hoe, water, and coddle the seedlings until they finally grow into food-bearing plants. Then it takes more time to pick the items for preserving. Before your food is in the jar or freezer, you will have devoted several hundred hours to it if you have gardened extensively.

Another advantage is the certainty of an adequate food supply. You can do all of the necessary work for a fine garden only to have a late frost, or early frost in the autumn, insects, animals, drought, or blights destroy your crops. If you drive to the store or market, you are guaranteed, except in the most unusual circumstances, that much of the food you desire will be available.

Buying also has certain conveniences that are not always a part of food growing. The bought vegetables are usually picked carefully, and only high-quality foods are offered for sale. Many times the foods are washed, shelled, or shucked and basically ready for use.

Finally, you may be able to save money. If a farmer has an overabundance of a crop, he might sell at a very low price rather than risk having the food spoil. In such cases you can often buy squash, for instance, for less than a dollar a bushel. Green beans can be bought for $.20 a pound, and tomatoes are sometimes available at $.10 a pound.

If you watch the produce counter at the supermarkets you will notice reduced prices on some containers of food which is perilously close to passing its peak of ripeness. If you make the grocer an offer of $5.00 or $10.00 for his entire supply, he might sell you several bushels of the item.

In the summer of 1982, for instance, small farmers near our home actually gave away huge amounts of fresh vegetables rather than watch it rot in the fields. Many people who garden always plant more than they possibly need, and they will share their surplus readily with others. Often the vegetables are given away, or you may be able to pay a nominal amount and pick your own vegetables or fruit.

The pay-to-pick system is especially common among the growers of strawberries, peaches, apples, and vegetable crops that are tenuous and expensive to pick commercially. Many vineyard owners also offer opportunities to pick your own grapes at reduced prices (Figs. 4-1).

Disadvantages

There are disadvantages that may nullify any gains you might have foreseen. The foods may be the products of forced growth through high-powered chemical fertilization, and the vegetable itself may be pithy and lacking in quality as a result of its too-rapid growth. Many times foods that grow too fast will split or crack, and the fissure becomes an invitation to insects or decay-causing enzyme action.

The widespread use of pesticides must also be considered. Many pesticides are highly dangerous to man, and even repeated washing may not remove all vestiges of the poison. Leafy vegetables such as lettuce and cabbage are especially hard to rid of all traces of pesticides. Many environmentalists warn of the dangers of many sprays that are used on large farm operations. These same foods, however, are often sold to supermarkets, and it could be argued that, if the foods are safe from the produce counter, they are safe from the wholesaler.

Often, too, the quality of the bought food is inferior to that which is picked fresh from the garden. Many foods, such as tomatoes, apples, and peaches, are picked slightly unripe. There is an obvious reason. The food must be picked, hauled to central headquarters, packed, shipped to the grocery store, and then prepared for sale. It may remain on the shelf for several days before it is bought and used. By the time the consumer gets the food, it has the appearance of ripe, vine-ready produce, but it lacks the taste and nutritional value of vine-ripened food.

If the producers wait until the food is ripe before it is picked, it will start to decay before it

Fig. 4-1. Many orchard, vineyard, or farm owners will allow people to pick for a fee. Many vineyard owners will permit free picking after the grape-picking machine has been through the vineyards thoroughly.

reaches the supermarket shelves. In either case you may be buying edibles that are decidedly inferior to fresh food.

Some foods, such as strawberries, will not ripen after picking, so often you must buy and use unripe strawberries or do without. Naturally, you cannot expect superior quality in canned or frozen foods if you start with substandard foods.

Just as your own garden may have a crop failure, so may the commercial growers experience damage to crops, or at times, complete destruction. The summer of 1982, for instance, was a summer of shortages in the apple and peach crops in the South. A late spring freeze and snow storm virtually obliterated the peach crop, and some orchard owners reported that they did not harvest one single peach or apple.

Vegetable crops can also be damaged on a large scale. Uncertainty of the crop can become a problem, therefore, and if you depended upon market suppliers, you may be left lacking in the pantry area.

Finally, while you can at times find lower prices, the likelihood is much greater that you will have to pay much higher prices for most foods. If you decide to can tomatoes, and you find them at the stores for $.39 per pound, you may find yourself paying over $20.00 for a bushel.

The final price of some items is, after all, nearly shocking. You may notice that a vegetable spaghetti squash sells for only $.49 a pound, but when you realize that four of the squash may cost you nearly $10.00, you see quickly how expensive purchased food can be.

From a general approach, one of the best reasons for preserving your own food is to avoid the high costs of groceries, and bargains in food are becoming increasingly rare.

GROWING YOUR OWN FOOD

There are decided advantages and disadvantages to growing your own fruit and vegetables.

Advantages

Among the favorable aspects of home growing is that you have your own choice of varieties of a certain food. You may prefer the taste of pole beans to string beans, Better Boy tomatoes to Rutgers tomatoes, Monroe peaches to Georgia Belles. If you grow your own, you can plant the types you like best, and there are literally hundreds of varieties of vegetables and fruits.

Another significant factor is that you have control over the pesticides used. If Japanese beetles attack your garden, you can put out harmless—to humans—beetle traps that are baited with fruit pulp rather than poisons. You may hand-pick your potato beetles rather than spray the plants.

Even if you spray, you know when you sprayed, how much you sprayed, and what type of spray was used. If you spray on Monday, there is a huge downpour of rain on Tuesday, and you pick the crop on Thursday, the spray will be negligible.

You are guaranteed, too, of garden-fresh produce. You can pick your ripe corn at eight o'clock in the morning, have it shucked and ready for blanching by ten o'clock, and in the freezer shortly after lunch. You are assured of food that was selected at the exact peak of ripeness and flavor, and preserved before that quality was lost (Fig. 4-2).

If you pick your green beans on Monday and leave them in bags on the porch until Friday, however, you have lost the freshness that you want and need. Food should be processed on the day it

Fig. 4-2. If you do your own gardening, you can pick corn while the dew is still on it and have it in the freezer or canner within two hours, or slightly more. It is at its best when preserved under these rapid-paced conditions.

was picked, if possible within three hours of harvesting.

To many people, the greatest advantage of growing your own food is the vine-ripened or tree-ripened taste. In truth, there is little that can surpass the taste of a peach picked at the moment of its peak ripeness. The taste of fresh beans, okra, corn, squash, and dozens of other home-garden foods is on the same level.

Not at all insignificant to the gardener are the joy, pleasure, feeling of well-being, and aesthetic satisfactions of watching and helping the garden in its process of producing visually and gastronomically delightful crops.

The healthful elements involved with the physical exercise should be considered also, and the recreational attitudes are of value to many people whose gardens provide them with their major sources of real exercise.

Disadvantages

It must be admitted that there are disadvantages to growing your own food. As mentioned earlier, there is always the uncertainty of the growing season and the yield of the garden. It is most demoralizing to watch a patch of corn flourish until just before the ears are starting to develop and then have a prolonged drought arrest the growth and ruin the patch completely.

The reverse may also be true. A week-long rainy spell can cause ripe and ready beans to sprout and decay, corn to bolt, melons to split and rot, and the insect population of the garden to multiply discouragingly. You cannot pick the ripe food because the garden is a morass rather than a beauty spot, and you lose large amounts of what would have been a cornucopia.

Gardening harvests or cultivations, too, seem to coincide severely with your spare time or business trips. A beautiful garden when you leave for the vacation trip can look like a small jungle when you return, or the ready harvest is past its peak and badly damaged by the delay. If you don't have extra time that you can spare for the garden, you may be fighting a losing battle.

You must also consider the costs of seeds and fertilizer, which can be considerable if you garden extensively. Many seeds are products of hybridizations, so you cannot save the seeds from an especially delicious melon, tomato, or other food and plant them the following year to get the same type of food. Silver queen corn is amazingly delicious, but the kernels from this season's crop will not produce silver queen corn the following year. If you plant seeds from a hybrid fruit or vegetable, however, you will get something edible the next year if you have a good growing season, and often the type of produce is a pleasant surprise.

The last disadvantage is the need for space. If you agree to share a garden spot with neighbors or friends, you may wind up in disagreements; if you must rent a plot, much of your savings will go to pay for the use of the land; if you live in an apartment or condominium or on a small lot, you may not be able to spare the space.

All things considered, though, the advantages of growing your own fruits and vegetables seem to outweigh the negative aspects by a tremendous margin, and we heartily recommend growing your own if at all possible and reasonable.

RAISING YOUR OWN MEAT AND POULTRY

The problems in raising your own meat, poultry, and fish become acute very quickly. While there are many advantages, the disadvantages loom menacingly for many if not most people. Unless you are in a particularly advantageous condition, we really do not recommend growing your own meat.

The major advantages of raising your own meat are quality control, possible monetary benefits, and convenience, plus assurance of a meat supply.

The disadvantages are the often high costs of raising beef and pork, the need for time-consuming personal attention, space requirements, and the uncertainty of the results.

Raising Beef and Pork

If you are raising beef and pork, you can use high-grade feeds, provide excellent pastures, and then pen and corn-feed the stock for the last six weeks before slaughter. You will have, in all probability, good meat, assuming that you started with good stock and have favorable climate, a good water supply, and an absence of calamity. You will also have the comfort of knowing that the meat has not been contaminated or exposed to harmful additives.

You might be able to save money, but there is much doubt in the minds of many on this point. If you have a good and inexpensive food supply, you may be able to realize a profit or a savings.

The cost of a whole beef or of a half or a quarter of a beef is considerably higher than the price of pork, and you might be more likely to save money on beef. Even if you don't save appreciably, there is the convenience and comfort of having the meat ready for slaughter or frozen and ready to eat.

The disadvantages of raising meat seem significant. For one thing, the high cost of food and the cost of buying the young stock can be great. During the long winter months you will need to feed the animals corn, grains of other types, hay, or other winter foods that you must grow or buy.

Perhaps the demands on your spare time are the most important considerations, however. Unlike a garden, orchard, or lake full of fish, it is not always possible for you to leave your livestock for prolonged periods of time unless the weather is such that they can eat in the pasture and drink from a certain water supply. If you must haul feed to the animals and water them by artificial means, you can become anchored to your immediate surroundings.

Then, too, you need space. In most instances, unless you have several acres of land, you should

Fig. 4-3. Chickens can be kept in a rather small area, and they can be fed corn and other foods from the garden and on dried corn in winter. In turn, the chickens produce eggs and fresh poultry meat.

not consider raising beef. Hogs require less space, of course, but if they are penned they will require daily attention.

The uncertainty of the results can be important to you, also. A slaughtered hog can spoil, just as a beef can, unless you have controlled climate conditions. You can pay someone to do the work for you, but you lose some of the savings by this expenditure.

If the meat is undamaged, you will not be assured that the hams and shoulders will be lean and heavy any more than you can know that the steaks and roasts will be lean, tender, and flavorful. Cured hams, too, can spoil, as can other parts of the hog that are cured.

Many people feel that buying the whole or half a hog or beef, may be the best economical angle, especially if you buy from reputable people in your area.

You might wish to consider that, on a whole beef or hog, about half of the carcass is bone, fat, and other waste. If this estimate is true, then a beef that you paid $1.49 a pound for is in reality costing you $2.98 a pound.

As we said earlier, there are numerous arguments that can be advanced on the subject of producing meat at home, and we feel that it is economically feasible to wait until the meat markets hold sales and then stock the freezer. If you happen to have the land, the food supplies, and the other necessities, then of course raising your own beef or pork is to your distinct advantage much of the time.

Raising Chickens

Raising chickens is another matter. You can raise a small flock of chickens in a very small space, and the food does not cost a great deal. You can grow a patch of feed corn and grind it in a home grinder for the smaller chickens and feed the whole kernels to the larger ones. Table and garden scraps will round out the food supply in many cases. Even if you must buy commercially marketed feed, the cost is not great, since chickens do not have the ravenous appetites of hogs.

It might be argued that neither do chickens produce the meat that a hog will. When you consider the space requirements, the fact that chickens can run free in many instances and areas, and that the loss of one chicken is not a catastrophe, you might agree that it may be more practical to raise birds and get fresh eggs as well as meat for the table. Chickens, too, can produce many chicks from one setting, and they need very little care.

If you decide that you'd prefer to let the butcher supply your meat needs, you can still save a great deal of money by buying at special times of the year and when there is a drop in the market price or a sale. You can preserve the meat and the eggs easily and handily at home.

5

How Much Should You Grow?

The amount of food you grow depends, of course, on many factors, such as the amount of land you have ready access to, the number of people to feed, and the amount of time and energy you can devote to your crops.

We can, however, establish some basic guidelines from personal experience and the experiences of others, and from information provided by nutrition experts.

This chapter will deal primarily with one-family food supplies of fruits, nuts, berries, and vegetables. Your own personal appetites will determine which of the crops and how much of them you will wish to grow.

FRUITS

Fortunately, fruit trees are rather inexpensive to buy, considering the vast amount of fruit they will yield over a period of years. They are relatively easy to care for, hardy enough to survive all moderate and some rather harsh climates, and they offer great beauty in addition to their fruit. The trees do not require immense amounts of space in which to grow, and bees take care of the pollinating.

Happily, fruit trees also do not take long to start bearing. If you wish, you can learn very easily to graft your own trees so that you can enjoy a wide variety of apples, for instance, in a small plot of land or even on one tree. Apple grafting is very easy and it can be done in a short time.

A one-family orchard in a typical cross-sectional climate might well include apples, peaches, pears, plums, cherries, and grapes. Many climates preclude the growing of oranges, grapefruits, and other citrus fruits.

We recommend at least three types of apples: a summer apple, a red fall apple, and a yellow fall apple. Summer apples, and there are many of them to choose from, ripen as early as June in most parts of the country, while others ripen in July, August, September, and October. If you have the space, ideally the best choices would be for five trees that would give you continuous apples from early summer into early autumn.

If space permits, too, we recommend either the standard size of the tree or the semi-dwarf. Dwarf fruit trees were very popular several years ago, and they still have a definite place in our environment. The larger trees, however, have more apple-producing potential.

If your space is very limited, however, you will find that the dwarf trees, which can be grown in an amazingly small area, are perhaps perfect for your needs. If you have a yard area that is 15 by 25 feet—not much larger than a huge room in a house—you can grow five or six dwarf trees without crowding.

Remember that your apple trees, like many fruit trees, will need a pollinator, so ask your orchard supplier to recommend trees that cross-pollinate well. Some trees cannot cross-pollinate others, and some can cross-pollinate only selected types of trees.

Cherry trees that are self pollinating can be purchased from nurseries around the nation. The younger the tree when you buy it the cheaper it will be and the greater its chances of strong survival; the smaller trees adapt to new soil better than those with well-developed root systems and established growth in its parent soil. One full-sized tree will provide you with enough cherries for your family.

If you have space enough, you might want two trees or more, but certainly you would want one sweet cherry and one sour variety for pies.

Cherries ripen in the mid-American climate zones in June and July generally. At cherry-picking time you can enjoy them and can, or freeze, or dry cherries for the winter. Cherry trees can be bought in dwarf as well as standard sizes.

Another tree you will surely want is the peach tree. Peaches ripen in June and up through July or August, depending on the climate and tree variety. The trees can be bought in standard or dwarf varieties.

You will certainly want one early-bearing tree and one late-bearing peach tree, so that all of your peaches won't be ripe at the same time. Peaches will be harvested from your new trees within 2 or 3 years after you set them out.

You can also purchase nursery stock for growing your own nectarines, apricots, and many other delightful fruits. If you can spare the space, we strongly recommend that you have as many types of fruit trees as you can use.

You will find that fruit trees can be grown fairly easily, but you will probably need to spray at least twice a year to protect your trees from insects.

Your orchard can be rounded out with a pear tree or two, and at one corner of your yard, or somewhere near the house, you may want two or three grape vines. You can buy white, blue, purple, and golden varieties, and you can buy grapes with seeds and grapes without seeds. Grapes ripen anywhere from July through September, and you will want to spread the harvest for as long as possible.

NUT GROVES

Nut trees, unfortunately, often require a very long time to bear their first crop. Pecan trees are notoriously slow, but other nuts are faster than either pecans or walnuts. You can start your nut grove by setting out trees or bushes of several types, enjoying the early varieties while you wait for the later types to mature.

We recommend two pecan trees, two filbert bushes, one or more almond trees, and some English walnuts. While the pecans may take as long as 12 years to bear, the almonds and filberts (often called hazelnuts) will bear in 2 or 3 years. The English walnut bears in 4 to 6 years.

You may not feel that the long wait for pecans or black walnuts is worth it. Since, however, the soil occupied by the trees might not be used for anything but lawn, you can still have your lawn under the trees, and the trees themselves provide beauty and excellent shade even before they start to bear (Fig, 5-1).

BERRY PATCHES

Among the berries and similar foods that you can grow with ease are strawberries, blackberries, blueberries, currants, raspberries, and elderberries.

For strawberries we suggest small patches about 5 feet wide and 15 feet long. If you can spare

Fig. 5-1. Walnut trees provide superior-tasting nuts as well as shade and decoration for many years.

the room, allow a walk space of about 2 feet, and set out another patch beside the first one. The best berries of all, we think, are the wild strawberries that will grow virtually anywhere and have an astonishingly great flavor. Wild strawberries make superb pies, tarts, jams, wines, and other delectables.

Among the cultivated strawberries are the so-called every-bearing plants that will provide berries from May through a killing frost period in the fall.

There are thornless or briarless blackberries that can be purchased from nurseries, but the old-fashioned types of blackberries remain our particular favorites. They will grow virtually anywhere, we have found that if you set out several plants in a patch 5 feet wide and 20 feet long below the garden spot, you will have excellent berries. The rainwater washes fertilizer from the garden to the berry patch.

Raspberries can be grown much like blackberries, and if you can't afford to have both, we suggest sticking with blackberries, since they grow so vigorously and produce so prolifically in all types of soils.

Raspberries and blackberries will bear during the first season, and blackberries are especially disease-resistant. Several varieties of blueberries are available, and they are among the most delicious berries that can be grown anywhere. You will need several bushes for cross-pollination and for abundant crops. Blueberries ripen as early as June and as late as September.

You may not find boysenberries, gooseberries, and elderberries in stores, since many of these berries do not hold up well during shipping, so if you want to enjoy them you have to grow your own. These berries are delicious, hardy, and require very little growing space. They do well as border growths and are generally disease-resistant.

VEGETABLES

If you must give space your primary consideration, then by all means reserve as much of your property as you can for your vegetable garden, because from this garden will come your staple foods for the entire year.

Fruits and berries are great for desserts, but you need potatoes, corn, beans, and other vegetables for your main meals two times a day if not three.

How Much to Grow

The amount you grow depends on the size and appetites of your family; so we will make suggestions for the needs of a typical family of four. Since, however, we have no way of knowing how rich your soil is or how good growing conditions will be, the amounts recommended are subject to question on all points.

You will probably discover that three or four long rows of potatoes—rows 60 feet long, for example—will take care of your potato needs for the year. Your yield will be much greater if you plant your potato seeds or eye-sets in a shallow furrow and then, as the plants grow, keep mounding soil around the plants until you have a hill a foot or so high. The potato plant will continue to form potatoes as long as the plant area is covered by soil. You can double or triple your harvest by mounding regularly.

We also discovered that if you plant your potatoes in the fall and cover them thoroughly with a thickness of dead and ready to decompose leaves your potatoes will have an excellent early start the following spring, and you will have an earlier and better crop.

Another staple food is the string bean, and most people make the mistake—if an overproduction is ever a mistake in gardening—of planting too many. One row 50 feet long will provide more than enough beans for fresh eating and for preserving. If you prepare string or a trellis for your beans to grow on, they will produce more, you can pick them easier, and there will be less damage from insects and rot.

If the row does well, you can pick a large grocery bag of beans every 6 or 8 feet, and from this bag of beans you get at least 7 tightly packed quarts of beans for canning.

It may be power of suggestion, but it seems that the more often you pick the beans, the more the plants will produce. So plan to pick the first fourth of the row on Monday, for instance, the second fourth on Tuesday, and so on. When you have completed the row, start again, and repeat the cycle.

One of the most prolific plants in the garden is the squash. Typically, three or four plants will provide all the squash you can use, but if you are concerned that there might be insect damage or that a plant or two might die, set out ten hills. Then be prepared to cook, eat raw, can, freeze, dry, and give away squash for the majority of the summer.

One row of okra will provide an abundance of food for your family, and the yield will begin in July and last into frost typically. The okra plants do not require much room to grow, and, like beans, they seem to thrive on frequent picking.

The amount of corn you should grow depends on how often you plan to eat corn during the year. Each cornstalk will produce about three to four ears of corn. If you plant 4 rows of corn with 40 plants to the row, you will harvest about 480 ears of corn. If you decide that each member of the family will eat 11 ears of corn every third day, you will need about 488 ears for the entire year. So plant more or less, according to the needs of your family (Fig. 5-2).

Like squash, cucumbers proliferate amazingly, and you will need only 8 or 10 hills of cucumbers to provide you with plenty of fresh cukes and several types of pickles for the winter.

One short row each of asparagus, beets,

Fig. 5-2. Plant rows of corn side by side for best pollination. If you intend to plant only one row, plant two half rows instead, and make the rows parallel. A good stand of corn will produce three or more ears per plant.

turnips, collards, radishes, carrots, spinach, chard, and similar foods will meet your needs.

Tomatoes are difficult to predict. One year we set out two plants, one on each side of the back door, and we daily dumped eggshells, tea bags, coffee grounds, bean hulls, peanut shells, and other compost-making materials around the plants. From one plant we picked 149 enormous and perfect tomatoes, and from the other we harvested 137 similar tomatoes. From these two plants we were able to eat fresh tomatoes all summer and also can more than 25 quarts for the winter. (Fig. 5- 3).

We have never since had such luck, and we suggest that you put out 25 tomato plants to ensure that you have an adequate supply.

Vegetable spaghetti plants also produce heavy yields, and three or four plants should take care of all your needs (Fig. 5-4).

Two long rows of peanuts should give you a large harvest, and a small melon patch will provide you with fresh watermelons, honeydews, and cantaloupes for the summer and for fall drying. You may want eight plants of each. If you have a good season, though, five plants might meet your needs.

You might also want a row of cabbage, a short row of lettuce, a row of onions, and a long row or two of early peas.

Conserving Space

If space is a problem, you can conserve your garden room by doubling up or by planning a second and even a third garden. You can have most of your first garden already harvested by mid-July, and you can set out a second planting.

You can save space by planting string beans in the corn row and then letting the beans climb the corn stalks. You can do the same with pinto beans, pole beans, and even cucumbers. You can let cucumbers or beans run up the fence that borders the garden.

Onion sets, radishes, and other early crops can be planted between the cabbage or potato plants. While the cabbage plants are still too small to require much space, the onions are reaching maturity, and by the time the cabbages need more room, the onions have been harvested, as have been the radishes and other smaller items.

You can plant pumpkins and melons among the corn, too, and the melons and pumpkin plants can spread over the corn patch and help keep weed growth down.

There are bush cucumbers and bush beans that take less space then the regular spreading varieties. Herbs can be grown in your flower beds rather than in your garden because they are attractive and take very little space.

Fig. 5-3. Tomatoes that do not ripen fully on the vine can be stored on a sunny shelf where they will continue to ripen to their fullest stages.

Fig. 5-4. Vegetable spaghetti plants flourish anywhere that squash grows well, and one large plant will feed four or more persons.

You can set out a lettuce bed in the very early spring and then use the space for other crops during the heat of the summer when lettuce grows poorly. In the fall you can reset the lettuce bed and have fresh lettuce well into the winter simply by covering the patch or bed with a sheet of plastic.

You can set out many vegetables, such as cauliflower, brussels sprouts, and broccoli, very early or very late so they do not conflict with other crops and compete for space.

Rhubarb can grow over in a small corner of the garden, since it takes little space and produces abundantly, and since many families do not want to eat a large number of rhubarb pies, which are delicious but not as good as apple or cherry cobblers.

In the blackberry patch wild strawberries can grow. They do not ripen at the same time; so you will have picked all of the strawberries by the time the blackberries are starting to ripen. Wild dewberries can grow along with the blackberries, too.

There are tiered structures you can make or buy for growing strawberries and cucumbers, and you can grow a very large amount of these crops in an incredibly small area.

Space limitations can be handled intelligently, and you will be astonished to learn just how much can be grown in a 10-by-20-foot garden plot if you mulch, plant closer, and stagger the crops.

6

Preserving Fruits and Tomatoes

We have chosen fruits and tomatoes as the starting points in home food preservation, not because they are any easier, necessarily, nor because they are most nutritious. We have chosen them because they are high-acid, therefore much safer for the beginner and, almost as important, they look so great on the shelf.

So fruits and tomatoes are discussed first, partially because they are so aesthetically enjoyable to prepare. The first law of any home food preservation, a law to be restated for emphasis, is to start with foods that are in the peak of their ripeness, that are sound, and that are free of any types of structural weakness.

SELECTION

When you are picking your own foods, you can later take a few minutes to separate the ones that are well suited for preservation from those that should be discarded or eaten immediately. When you are buying these foods, you do not want to waste your time and money on fruits that cannot be canned, frozen, or dried effectively.

Some basic observations may be helpful. When you are selecting anything to be preserved, there are a few easy and commonsensical means of doing so. First, use the visual approach. Pick up the fruit or vegetables, turn it over, and examine all exterior areas to be sure there are no insect spots, moldy areas, or dark sections that indicate the beginning of decay.

If you are buying English peas or sugar peas, even though the hulls or shells are fat and healthy looking, you should hold the pea pod up to the light so you can see the outline of the peas. You may be surprised to find that, if the peas were picked prematurely, the peas are actually tiny. The outer hull develops first, and it is large enough to allow the peas to have growth room. If the pea is left on the plant the individual peas will fill the cavities (Fig. 6-1). If picked too early, they cannot develop.

If you are buying corn, particularly corn with the shucks still on, pull back the shucks or leaves at the top end and make a quick visual examination to see if worms or borers or other pests have attacked. One small worm, in addition to being a

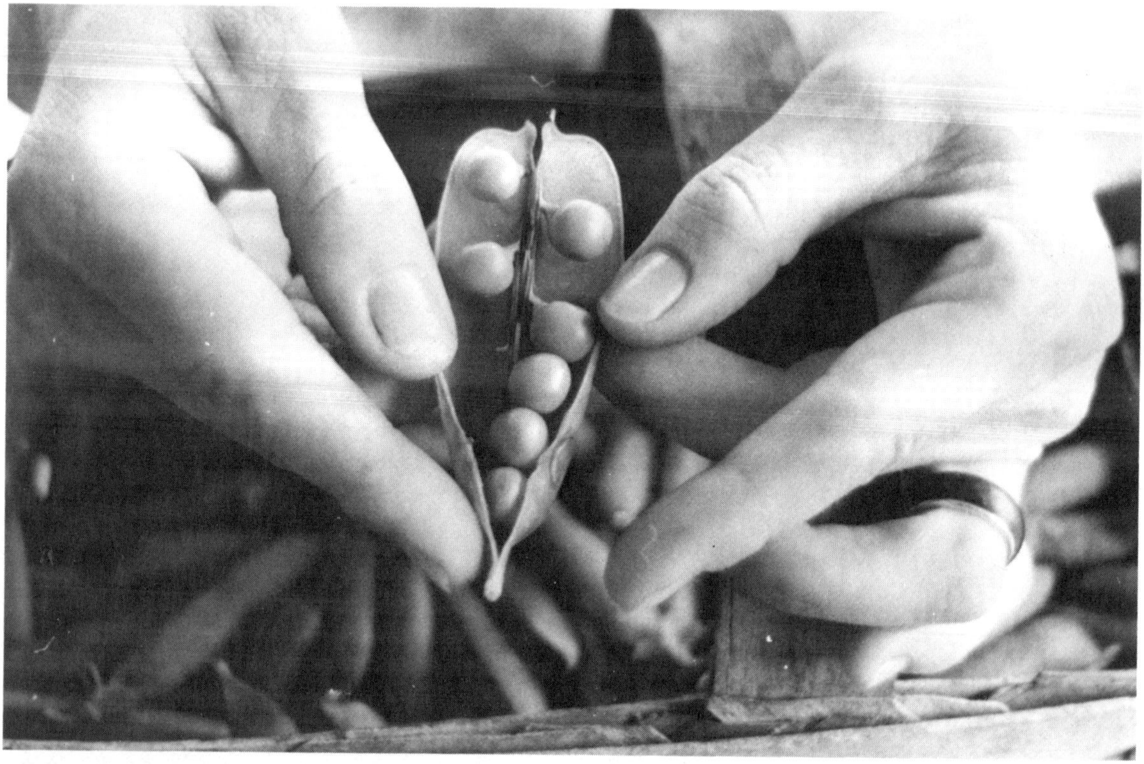

Fig. 6-1. Wait to pick peas until they have a chance to develop. A fully ripened pea will be large and round, and there will be six to nine healthy peas in one pod.

distasteful part of your purchase, can destroy an ear of corn and still leave it looking healthy and terrific on the outside. Look too for tiny holes in the shucks, especially in the bottom third of the ear. Tiny insects eat their way into the corn and then proceed to ruin the bottom part of the ear.

With tomatoes, check to see if it is totally ripe. It should be a deep red or yellow, generally, although there are some pink tomatoes available. The color should be uniform from the blossom end to the stem end. If the tomato is still green around the stem end, it was picked before peak ripeness, and you will have a hard and green interior section that does not can well. You can put the tomato on a shelf that receives sunshine and leave it for a day or so, and it will continue ripening. If, however, you have planned your day's activities around canning or freezing, you may not be able to wait for the fruit to ripen.

Tomatoes should be smooth, free of lumps, bumps, humps, and nodes. There should not be any wrinkled or shriveled places, and the tomato should not pucker at the stem end. It should not, of course, be cracked open or have tiny splits in the skin or peel.

Fruits are sometimes difficult to select visually. An apple turns red or yellow sometimes three weeks before it is fully ripe, so color alone is not adequate evidence. You should look to see, of course, that the apple is sound and colored well, but you might need to try other tests, too. You can use your thumbnail for a quick test; if your thumbnail penetrates the peeling and the flesh of the apple with a crisp crackling sound, the apple is ready to be used. If you cannot make a small indentation with your thumb, the apple is still too green. If the flesh is soft or spongy, or if the peeling is slightly wrinkled, the apple is past its peak of ripeness.

Peaches are very difficult to test. If the peach—or for that matter most fruits—had been allowed to fall from the tree, it will be soft and ripe, perhaps, where the impact with the ground bruised the flesh, but it might be hard and green in other areas. Such peaches are not ready for either eating or preserving.

You can hold the peach in your hand and apply light pressure with thumb and fingers. If you feel the peach flesh give slightly, and if the coloring is right, the peach is perfect. But the slight pressure should produce results all around the peach, not just on one side.

Another excellent test is to pinch a tiny bit of the peeling off and then see if you pull away large sections of peeling easily. If you can, the peach is ready.

Plums and apricots should be firm, yet they should yield slightly to light pressure.

If you are picking your own fruits or tomatoes, you can tell which fruits are ready for use if, when you start to pick them, they fall away from the limb or plant without any real resistance. If you have to pull or tug, the fruit is not ready to be used.

Green beans should snap crisply under slight pressure, and the break should be sharp and clean. If the bean bends rather than breaks it is too mature.

If you are planning to work with melons, a few simple tests will help you to select the best possible foods to use. With cantaloupes, a perfect test is to disregard the color and other visual or kinesthetic tests and simply hold the melon up to your nose and take a good whiff. If the melon smells strongly sweet, even through the rind, this is the one you want. If it has no sweet odor or only a faintly sweet smell, pass it by and choose another.

With watermelons, place both hands on the top of the melon and push down with slight firmness, enough to cause the melon to give or yield, and listen intently for a crisp crackling sound from the rind. It will be a faint but clearly discernible sound. If you hear it, the melon is likely to be a sound and ripe one.

One other test is also simple: after you examine the food, ask yourself if you would pick this particular fruit or these vegetables for your next meal. Are they ripe enough yet not too over ripe for the meal? If they don't suit you for the present eating, they won't be ideal for later eating either.

What difference do the stages of ripeness make? If the food is underripe, it has not developed all of its flavor; if overripe, it has lost much of its taste, and enough enzyme and bacterial action has started to perhaps cause the food to spoil.

PREPARATION

Preparation for food preservation has many implications. You should check your supplies to see that you have all of the ingredients, spices, jars, lids, freezing bags, and that all other items of equipment are cleaned and sterilized.

Wash your food thoroughly, using a soft brush if necessary, to get all dirt and grit off, and then drain until no water is left standing on the food.

Foods like blackberries are ready to be processed once they are washed. Cap strawberries after washing; string and break green beans. Check all food to determine that all bad spots have been cut out. Some foods, like peas, must be shelled. You will need to wash carefully not only to remove dirt but also to rid the food of spray residual presence and to eliminate wild yeast.

Now, sort foods, when appropriate, according to size. Some individual fruit may be small enough for you to can them whole, while others will need to be sliced or at least quartered.

You are now ready to process.

WATER-BATH CANNING

No matter how you plan to process your fruits or tomatoes, there are a few obvious rules to follow. First, read all directions before you actually start work. If you will need to make a trip to the store or prepare equipment, do so before you immerse your food in boiling water or before it is even peeled and exposed to air and enzyme action.

The ways of processing discussed here will be water-bath canning—both raw-pack and hot-pack methods—pressure canning, drying—oven, solar, and dehydrator methods—freezing, and root cellar storage. The most basic method to many people is the water-bath method, and that is where we will start.

Equipment

The equipment necessary for water-bath canning, or *boiling water canning*, other than jars, lids, and necessary utensils, includes the boiling vessel and a rack. You do not necessarily have to have a water-bath canner if you have any sort of pail, vat, or huge pot large enough for you to submerge the jars in water and have an inch or so of room above the tops to allow the water to boil.

You will also need a rack of some type to keep the jar bottoms from being in too close contact with the source of heat. Old-fashioned methods included the use of a spoon or similar eating utensil in the bottom of the vessel so the jars could sit on it, at least partially.

Remember that the water-bath method is to be used only for high-acid foods, such as fruits and tomatoes. Neutral or low-acid foods canned in the water-bath method can be a tragic source of botulism. We discourage the use of the water-bath canner for low-acid foods at all times.

Once the food is prepared, fill the water-bath container with water and start it heating. While it is heating, you can pack jars in one of two ways: the hot pack or the raw pack methods.

Raw Packing

Raw packing means putting the food in the jar and then pouring either boiling water or a hot syrup over the food. First, pack the fruits or tomatoes into the jars as firmly and tightly as you can without crushing them. If you are canning small tomatoes, you may want to leave them whole. If you are processing peaches, you may want to halve them and pack so that the pit or seed cavity is always turned to the inside and therefore never visible to detract from the beauty of the canned fruit.

While many thousands of people have safely canned whole peaches, we discourage this method for one basic reason: peach seeds contain cyanic acid. If the seed remains intact, there is little chance of contamination. If, however, the peach seed somehow splits open after the processing, the acid can seep into the syrup and eventually into the flesh of the peaches, and the acid could cause serious illness or even death to the consumer.

After your fruit is packed, you can pour a syrup over it, or you can use juice that has already been extracted from the food. Let the juice or syrup have time to seep into all the nooks and crannies, and then make sure you have covered the food but that you have left at least 1/2 inch at the top to allow for expansion during the processing.

If you want a light syrup, which is used with many small and soft fruits, you can make the syrup by mixing one cup of sugar with three cups of water and then boiling until the sugar is completely dissolved. You can also use juice from the fruit and mix it with sugar and boil.

A medium syrup, which should be used with such fruits as apples, peaches, pears, and cherries, can be made by mixing one cup of water or juice with two cups of sugar and boiling until the sugar is dissolved.

A heavy syrup, to be used with extra-tart fruits, consists of one cup of water or juice and one cup of sugar. Boil until the sugar is dissolved.

When you are canning tomatoes, you can pour the tomato juice or water over the food before closing the jars.

Hot Packing

The ***hot-pack method*** is one in which the fruits are cooked for a specific time—a listing of times for various foods will be provided later in this chapter. Pack the hot, cooked food into jars, pour liquid from the cooking vessel over the food, and then put on the lids and rings.

While you are packing the food, your water-bath container should be heating, and it is a good idea to start another pot of water heating at the same time.

Processing

When the jars are packed and closed, the water in the processor should be close to boiling. In fact, temperatures of the water in the processor should be as close to the temperatures of the food and jars as possible. If the food and jars are near two hundred degrees and the water is at a much lower temperature, there is a possibility that the jars will crack when they are placed in the water. If the water in the processor is boiling and the jars and food have been allowed to cool too much, a similar danger of breaking exists.

The rings and lids can be in the second pot of water that is heating. You can take out the lids as you are ready for them by using tongs, a knife blade, or fork tines. Before you put the rings on, run your finger around the rim of the jar opening to be certain that there are no nicks or rough places that would keep the jars from sealing later.

When you are ready for the lids, place them so that the lid is forced down firmly against the top of the jar rim. Tighten the ring as firmly as you can get it without using any mechanical devices.

Using tongs to keep you from burning your fingers, lower the jars carefully and slowly into the water inside the processor. It is good to dip the bottom of the jar an inch or so into the water, wait several seconds, then lower it a few more inches, until the jar is fully immersed and the water covers the jar completely.

Wait until the processor water comes to a full boil, and then start timing the processing. Do not start timing before the water boils, and be careful to keep the water at a boiling temperature during the entire processing. If you need to add water during the processing, pour some from a second pot of water you started heating earlier. Always keep boiling water handy.

After the food has processed for the proper time, use the tongs to lift the jars out carefully and place them on a dry towel that has been folded several times. Do not set the jars on a table surface, since the heat may damage the table and the coolness of the table may cause the jars to break. Be sure the towel you use is dry, since a wet, cool towel or other cloth could cause jars to break.

Do not set the jars in a draft or in front of a fan, and do not place them close enough to the refrigerator that cold air from the opened refrigerator causes jars to break.

After all jars have had ample time to seal, test for correct sealing by a visual examination to see if the lid has curved downward, or press the lid with your fingertips. You can also tap on the lid with a fork or spoon, and if the sound you hear is a clear pinging noise, the jar is sealed. If it is a dull noise, the jar has not sealed.

Any jars that do not seal properly should be checked for problems by removing the ring and lid, making sure that no food is high enough to touch the lid, and that the rim of the jar is free of any irregularities. Then put on a new ring and lid and reprocess the jars that have not sealed.

If you have several more jars to pack, you can wait and run the unsealed jars all at the same time, but do not wait for more than an hour or so unless you protect the food from contamination by refrigerating it or by keeping it at a very high temperature. In the hot-pack method there is a problem of overcooking, so your best bet is to reprocess the unsealed jars immediately or serve the contents at the next meal.

OPEN-KETTLE CANNING

You can preserve a few foods, such as jellies, pickles, and other similar preparations through the open-kettle method, which consists of a few simple steps.

Clean and sterilize jars. Have rings and lids, also heated, ready. Fill the jars with the apple butter, peach butter, or pickles. Pour hot liquid, either juice or other preparation, over the food, if you need a liquid, to within 1/2 inch of the top. Place lids on jars, after first checking to see that the rims of jars are in good shape, and hand-tighten the rings. Let the jars seal, and the food will be ready to store as soon as it cools.

Do not attempt to use the open-kettle method for any types of vegetables, meats, or fruits, since there is always a danger of spoilage or botulism. Use this method only when the food has a sugar content or vinegar content great enough to prevent spoilage.

PRESSURE CANNING

Prepare the food for canning basically as you did for water-bath processing, but when you use a pressure canner put 2 quarts of water into the 21-quart canner and turn on the burner under the canner. You can put hot water into the sterilized jars until they are ready to use, and heat the lids and rings as you did in the water-bath method.

Pack the jars with the precooked or otherwise prepared food and fill jars within 1/2 inch of the top with whatever liquid you plan to use or with juices from the fruit or tomatoes. If you prefer, you can use the light, medium, or heavy syrup that you used in the water-bath method.

Use a clean knife, spatula, or spoon to work out any air bubbles that have been trapped in the jars, and then clean the rims of jars with a clean, damp cloth before you put on the lids and rings.

Stand the jars on the rack in the hot canner, and put on the pressure canner lid. Set the burner at maximum heat and wait until steam is clearly visible as it escapes from the vent pipe in the top of the canner lid.

If you have the manufacturer's booklet on how to operate your canner, follow the directions provided. In case the booklet is not available, here are the basic instructions.

After the steam starts to appear, reduce the heat just enough that steam will continue to jet out at a moderate rate, and wait 7 to 10 minutes so that the entire canner is filled with steam.

When all the air has been eliminated, place the pressure regulator on the vent pipe.

Within about two or three minutes, perhaps slightly longer, the automatic air vent should close because of the steam pressure being built up inside the canner. If for some reason the vent does not close on its own, tap the vent with a spoon or knife gently. You need to barely touch the plunger to cause it to seal.

After the vent has been closed, use maximum heat until the pressure has been elevated to the desired point, then reduce the heat. You may need to reduce the heat just before the pressure reaches the proper point in order to keep the canner from building excessive pressure.

As soon as the right pressure is reached, start timing the processing. Keep the pressure at the proper level by either raising or lowering the heat; you must watch the canner pressure carefully to see that there are no measurable fluctuations in pressure.

If for some reason you allow the pressure to fall below the required level, start your timing for processing over and proceed until the time for processing has been completed. When the proper time has elapsed, turn off the heat or move the canner to a cool burner or other surface.

Wait until the pressure drops to zero before you do anything to the canner. When the steam pressure drops to the proper point, the air vent plunger will drop to its original position, at which time you can take the pressure regulator from the vent pipe.

Even though the pressure has dropped, there will be a great deal of very hot steam still inside the canner, so do not open it for three or four minutes for cooling.

When you take the lid off, tilt it away from you so that the steam will not billow up into your face (Fig. 6-2).

As soon as you take the lid off, place a towel over the canner and wait for 2 or 3 more minutes before you take the jars from the canner.

When you are ready, grasp the rack or basket handles, using the towel edges as pot holders (Fig. 6-3). Lift the basket from the canner and place it on a thick dry towel (Fig. 6-4). Since the jars will still be boiling, don't set the basket in a draft or anywhere that will receive cold air. We recommend that you turn the air-conditioning unit off or the thermostat up so that the shock of the intensely hot jars hitting the cool air will not result in cracked jars. It can be terribly disheartening to spend so much time in canning the food only to have it ruined as you complete the final steps.

When the jars have cooled sufficiently, place them in a convenient space and allow them to set for 24 hours. You may now remove the rings and use them in other cannings. After 24 hours, check for proper sealing and, if all is well, store the jars.

Preparation and Canning Times

Below are basic preparations and canning times for a variety of fruits and for tomatoes. If the fruit you need to can isn't listed, find the fruit that is most similar to it and use the same times and pounds of pressure.

Fig. 6-2. Hold the lid so that escaping steam is directed away from your face and body when opening the pressure canner.

Fig. 6-3. Use potholders to protect your hands as you lift the jars, still in the rack, from the canner.

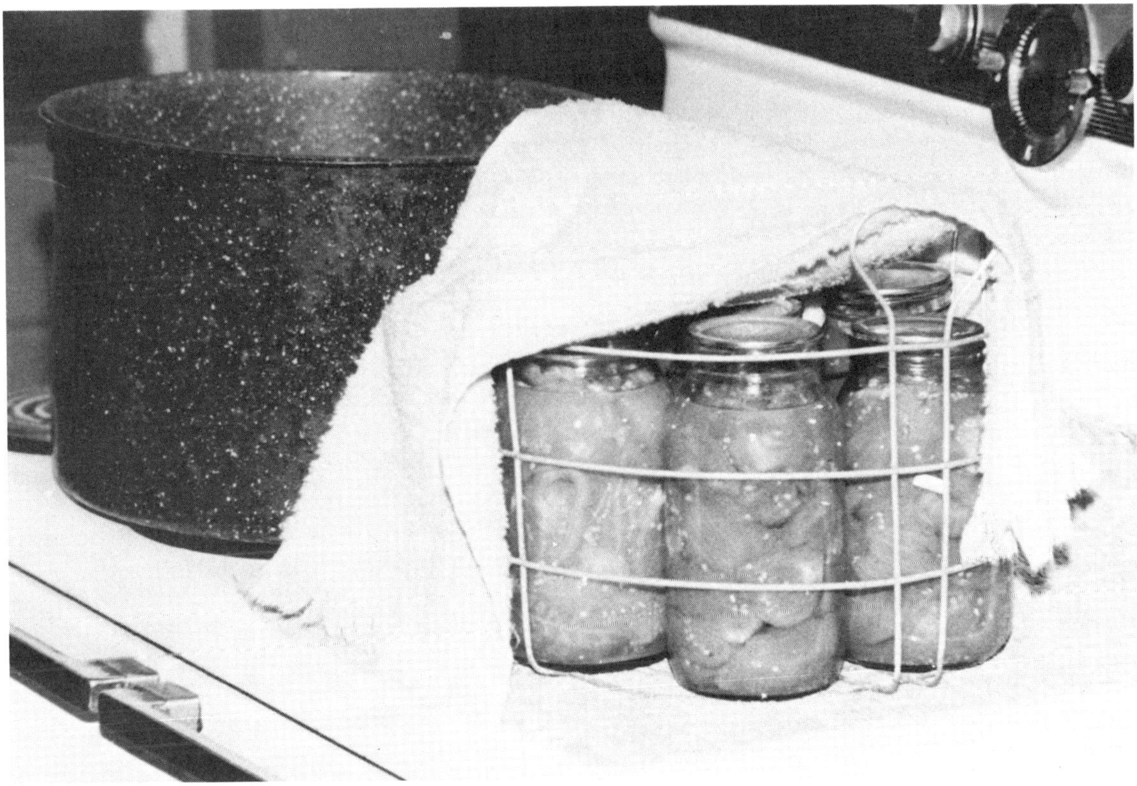

Fig. 6-4. Cover the hot jars as soon as you remove them from the canner.

Apples. Peel, core, and slice apples. Cook for 5 minutes, pack in jars, and pour either syrup or hot water over them. You can also cook the apples until the moisture has been evaporated, and you will have applesauce. You may want to press the cooked apples through a sieve or pack the applesauce after cooking. Process quarts of apples 15 minutes in a water bath or for 10 minutes at 5 pounds of pressure.

Apricots. Wash, cut in half, and remove pits from apricots. After you drain them, pack in jars and cover with boiling water or syrup. Process for 30 minutes in a water bath or for 10 minutes at 5 pounds of pressure.

Berries. Use firm berries and add eight or nine tablespoons of sugar per quart. Cook berries and sugar in a covered vessel until the boiling point is reached. Remove from heat and pack into sterilized jars. Cover with syrup or the juice from the cooking. Process in a water bath for 15 minutes or for 10 minutes in a pressure canner at 5 pounds of pressure.

Cherries. Most cherries can be pitted easily by holding the cherry between two fingers and gently squeezing until the pit pops out. If the cherries are a sour variety, they will need to be packed with a heavy syrup. Sweeter varieties require less sugar, and a medium syrup will suffice. You could instead use six to eight tablespoons of sugar for each quart of cherries, bring to a boil, and pack hot. Process in a water bath for 20 minutes or for 10 minutes in a pressure canner at five pounds of pressure.

Cranberries. After washing cranberries, remove the stems and drop berries into a boiling syrup for three minutes. Pack hot cranberries and syrup in quart or pint jars and process in a water bath for 10 minutes or in a pressure canner for 10 minutes at 5 pounds of pressure.

Figs. Cover thoroughly washed figs with fresh

water and bring to a boil. Then let stand for 3 minutes and drain, reserving liquid to make a thin syrup. Pack hot figs in jars and cover with the thin syrup. Add one tablespoon lemon juice to each quart or use a slice of fresh lemon, if preferred. Process in a water bath for 30 minutes or in a pressure canner for 10 minutes at 5 pounds of pressure.

Fruit Mixture. Small amounts of fruits left from other canning may be combined and heated in a medium syrup. Fill jars with warm mixture and process in a water bath the length of time necessary for the fruit requiring the longest processing or in a pressure canner for 8 minutes at 5 pounds of pressure.

Fruit Juices. Any kind of fruit juice left from canning should be heated to boiling and poured into jars. Process in a water bath for 10 minutes or in a pressure canner for 8 minutes at 5 pounds of pressure.

Juice from fresh fruit pulp may be obtained by crushing the fruit, heating slowly, and straining juice. Reheat juice before filling jars. Process in a water bath for 10 minutes or in a pressure canner for 8 minutes at 5 pounds of pressure.

Grapes. Grapes can be packed cold into jars and covered with boiling water or a medium boiling syrup, if more sweetener is needed for a sour variety of grape. Grapes may also be heated in a light syrup and packed hot. If a juicier type of grape is used, grapes may be packed in jars and crushed slightly until there is enough juice to cover grapes. Process in a water bath for 20 minutes or in a pressure canner for 8 minutes at 5 pounds of pressure.

Peaches. Peaches will peel easily if they are plunged into boiling water for about 1 minute, then dipped into cold water. Peel, slice or halve, and pit peaches. Place in a weak brine solution—1 to 2 teaspoons salt to each quart of water—to prevent discoloration of peaches until ready to use. Rinse, drain, and heat in a medium syrup for 2 to 3 minutes. Pack hot peaches into jars placing halves with cut side down. Cover with boiling syrup. Process in a water bath for 20 minutes or in a pressure canner for 8 minutes at 5 pounds of pressure.

Pears. Peel pears and cut into halves. Core and put pieces in a weak brine solution to prevent discoloration of pears until ready to use. Rinse and drain well. Simmer pears in a medium syrup until pieces lose firmness—about 2 minutes. Pack hot pears into jars and cover with boiling syrup. Process in a water bath for 25 minutes or in a pressure canner for 8 minutes at 5 pounds of pressure.

Plums. Whole plums may be canned if the skin of the plum is first pricked with a fork. You can also halve and pit them. Bring prepared plums to a boil in a medium syrup and cook 2 minutes. Pack hot plums into jars and cover with boiling syrup. Process in a water bath for 15 minutes or in a pressure canner for 8 minutes at 5 pounds of pressure.

Rhubarb. Remove leaves and ends from rhubarb stems. Cut stalks into lengths 1/2 to 1 inch. For each quart of pieces add 1/2 cup of sugar and slowly simmer in a vessel with enough water to cover the bottom. Bring to boiling and then pack into jars. Process in a water bath for 10 minutes or in a pressure canner for 5 minutes at 5 pounds of pressure.

Strawberries. Wash strawberries before capping. Measure 1/2 cup of sugar for each quart of berries. Bring slowly to a boil, stirring frequently to prevent sticking. Once mixture has reached boiling, remove it from heat and let stand at least 4 hours. Pack mixture into jars and process in a water bath for 15 minutes or in a pressure canner for 5 minutes at 5 pounds of pressure.

Tomatoes. Drop whole, unpeeled tomatoes into boiling water for 30 seconds to loosen skins. Then plunge tomatoes into cold water. Peel and remove stems and ends. Pack whole or quartered pieces—depending on tomato size—into jars and press down each piece. Juice will cover the pieces in the jars, or a hot juice made from less perfect tomatoes can be used to cover. Process in a water bath for 30 minutes or in a pressure canner for 10 minutes at 5 pounds of pressure (Fig. 6-5).

STORING CANNED FOODS

One fine way to store canned goods is to convert an old closet into a makeshift pantry. You can also

Fig. 6-5. While some tomatoes are in the boiling water, others are cooling so that you can peel them. Still others are nearby ready to be heated so they can be peeled easily.

find a corner or alcove in which to build shelves for storage. Or you can easily construct a simple cabinet in which to store your goods

In our own case we had a chimney for the old wood stove. When we put in the electric range we took out the chimney, built a floor into the space, and added walls around the spot and converted the chimney into a pantry that serves very well. We also used corner cabinets which are simple in construction in which to store and display our canned food. We did the same in the kitchen with a cabinet and display area (Fig. 6-6).

You can build easy shelves by using cement blocks and wide boards. The first rows of jars can sit on the floor, and the shelf boards will allow a clearance of 2 or 3 inches. The jars that won't fit on the floor can be stored on the makeshift shelves.

Many people pack the sealed jars back into the boxes the jars came in. After the box is filled, tape the flaps closed and mark what kind of food is in the case and the date canned.

Fill each box with the same kind of food when possible. Then you won't have to open half a dozen boxes in order to find the jars you want.

With the boxes clearly labeled, you can make a quick count of a particular type of food. Some people find it helpful to mark each time they remove a jar from the box, so you can see at a glance that you have nine, four, or one jar to last you until the spring garden starts to produce. You can also plan menus more effectively if you can conduct a rapid inventory.

When you stack the cases, stack all the same kinds of food in one stack. This way you won't have to transfer cases from one stack to another. You will want to use the oldest cans or jars first, so if you have some food from three years earlier, use this food before you open that which was canned last summer.

Do not stack boxes on damp floors. The

Fig. 6-6. A corner of a room can be converted into a pantry or shelf area. This pantry was once a chimney.

dampness will seep into the cardboard case and weaken the bottom so badly that when you try to lift the carton, the bottom will fall out and the jars will break.

AIR DRYING

To build an air dryer or solar dryer, you need six strips of wood about 2 inches wide, 1 or 2 inches thick, and 5 feet long. You also need six strips 30 inches long, and some screen wire, enough to cover the box you will make from the strips. Use four 2-by-2 sections of lumber 2 feet long for the corners.

Nail together a rectangle 5 feet long and 30 inches wide by 2 feet high. Staple or tack screen wire over all but one long side after you nail the final two strips down the length of the rectangle. Nail the strips so that there will be about 6 inches of air space between them.

You can stretch wire or cheesecloth between the two strips or rails and make a drying rack that you can suspend from nails driven in the 2-by-4 lengths that serve as corners for the dryer.

If you wish, you can make a door that fits into the side you have left open, or you can simply staple the section of screen wire over the opening when you are drying food.

When you prepare food for drying, you may blanch it for about five minutes. To blanch food, drop it into a pot of boiling water or steam it. The purpose of blanching is to stop enzyme action and break the tissues of the food down slightly so that the moisture content can escape easier and speed up the dehydration work. Steaming or blanching also helps the food to retain its color and nutrients.

Tomatoes

Dried tomatoes are very easy to work with, so we will start with this universally accepted food, which can be used in soups, casseroles, sauces, and numerous other dishes.

You need firm, fully ripe tomatoes which are free of all blemishes. You do not want to include a rotted spot in your dried food packages, since the decay is likely to contaminate the rest of the package. Cut out any sections that are inferior or which have been damaged.

One of the easiest methods for drying tomatoes is to slice the food the way you would if you were planning to make a tomato sandwich. The slices should be no more than 1/4 inch thick. You can also quarter the tomatoes if you prefer, but thicker wedges require more drying time.

After you have sliced the tomatoes, spread the slices on the drying screen or cheese cloth and place the rack in a dry, sunny place. If you have a concrete porch or other similar area where heat will be reflected well, you might want to place your solar dryer there. Another good place is against a white wall which will reflect heat well. The ideal place is a concrete floor with a white wall in the background.

Fill the drying racks with slices of tomatoes, and after the racks have been in the sun for 2 hours, perhaps slightly longer, turn the slices. The slices should be left in the sun for 8 to 10 hours, and you should turn the slices at least 2 more times for a total of four cycles.

If you prefer, you can dip the tomatoes in boiling water for a minute or two and then slip the skins or peels off before slicing. A supply of cherry tomatoes can be cut in half and dried like the slices.

When your food is dried thoroughly, slip it into an air-tight container and freeze it overnight, after which you can store it at room temperature for several months. It will keep as long as moisture does not permeate the flesh again.

Fruits

For apples, peaches, pears, and most other fruits, peel, blanch, and slice into 1/4-inch sections, then spread the fruit on the rack to dry. You may wish to dip the fruit into ascorbic acid before you dry it, or you may steam or blanch for several minutes—5 to 10 or not at all if you prefer.

To test for dryness, select a slice and cut it to see if juice or moisture oozes from the cut. If there is no moisture, and if the fruit feels leathery and limber, it is ready. General drying time is 8 to 10 hours in the sun.

You may dry the apples, peaches, and pears very quickly if you peel, core, and slice them into very thin sections and then turn the slices every hour or so.

Apricots should be pitted and cut in half, then steamed for about five minutes. Dip the halves into ascorbic acid if you prefer and then spread in the sun. Since apricots are thickly fleshed and highly moisture-laden, it may take two days of bright sunlight for the drying process to be completed.

You can dry cherries by pulling the stem from the cherries and squeezing the pits out. Much of the time if you exert gentle pressure with your thumb and forefinger and pull the stem with the other hand, the pit will come out at the time the stem is separated from the cherry. You can dry the cherries just as they are by spreading them on the rack and putting the rack in a sunny spot for ten hours or so.

When you are drying plums, cut the fruit in half and remove pits. Steam—this again is optional—for about 5 minutes or slightly longer, drain, and then spread to dry. Wait until the fruit is leathery and coarse-feeling. Drying time is usually at least 12 hours in the sun.

Generally speaking, drying time and blanching time depend largely on the size of the fruit and the thickness of the slices, as well as the temperature in the drying containers. Obviously, fruit will dry faster when the temperature is 95 degrees than 80 degrees. Cloudiness will also affect the drying time considerably.

For the above reasons, we do not attempt to prescribe an exact time for drying by the solar or air method. All fruits, even if they are of the same variety also do not always have the same moisture content. Much depends upon the growing season as well as the degree of ripeness and maturity of the fruit.

Grapes, like several other fruits, may be dried without any treatment at all. You may leave the stems on and spread the grapes on the drying rack and leave in the sun for 10 to 12 hours. At the end of the day the grapes should be shriveled, leathery, and pliable and spongy. When you can obtain them, be sure to use seedless grapes; in fact, grapes with seeds are seldom worth the trouble of drying unless you are totally unable to get any other kind and desperately want raisins.

Again, use your own judgment concerning the fruit and its dryness. Remember that you do not want the fruit to be totally dry and crisp; you want it pliant but rather tough to chew, and you should not be able to squeeze any moisture from it when the drying is completed.

With all fruits and tomatoes, store in an air-tight bag or glass container until you are ready to use the fruit.

You can also use dry bananas, pineapples, figs, dates, and berries of all kinds by the solar method. Slice bananas and place the slices on the rack and leave for 10 hours in bright sun. Slice pineapples very thin, dip into the ascorbic acid or blanch for 6 to 8 minutes before spreading. Slice large figs in half and dip into boiling water for one minute, then drain thoroughly and spread to dry. Berries should be washed and either spread to dry or dipped into boiling water for about 30 seconds and then drained. Dry and store these fruits as you did apples and peaches.

One of the most delightful of all dried foods is the watermelon. You can dry both meat and rinds and you can use the rind in making preserves and pickles.

Cut rings of watermelon, rind and all, 1/2 inch thick and then run your knife blade around the rind so that the meat will be freed (Fig. 6-7). You will then have a circle of red meat that can be seeded easily. After seeding, place the melon on the drying rack and leave in the bright sun for 12 hours. Turn the slices every 2 or 3 hours. If the drying has not been completed by the end of the day, cover the rack to prevent dew from adding moisture, and leave the rack in the sun for another full day, if necessary.

Dry cantaloupes, honeydews, and muskmelons the same way. When they are dried, they will be limber and have a coarse texture, somewhat like the pages of a newspaper or very thin leather.

OVEN DRYING

To oven-dry fruits and tomatoes, you should prepare the food the way you did for air drying. Generally, you should seed, pit, or core fruits with large seed or pit cavities. Blanch or steam, as you prefer, or dry without treatment.

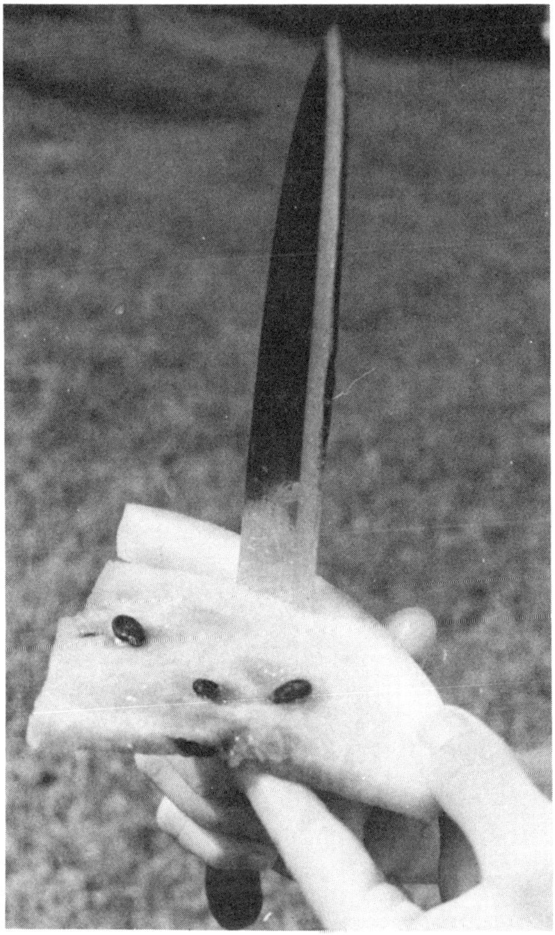

Fig. 6-7. Slice watermelon into wedges, trim rind, and remove seeds.

Some experimenters with drying insist that if you will mix 3 tablespoons of sodium bisulfite into 3 gallons of water and then soak your fruits and vegetables for 10 to 15 minutes before drying, the final product will be better. We have never found untreated food to be tough or in any way undesirable, however.

When you are ready to dry, you will need a rack of some type and a small fan. Set the oven heat on the lowest setting and crack the oven door about 3 or 4 inches. About 3 or 4 feet from the oven door set up a small fan and turn it on at the slowest speed. Let it run during the entire drying process.

Drying time will be much shorter than during solar or air drying, and the food will have a tendency to become a little crisper unless you watch it carefully, turn it regularly, and remove it from the oven as soon as it is adequately dried.

When you are oven drying watermelon and other melons in this fashion, you will find that the finished product is almost crispy. Since melons are over 80 percent water, you have removed over 4/5 of the total volume in size and moisture content. The result is that the sugar content becomes greater in proportion to size or volume, and the food becomes sweeter than fresh melons could ever be. In short, dehydrated melons must be among the most delicious foods available anywhere on earth.

If you overdry the watermelon or cantaloupe, you destroy the flavor, so be very alert during the last hour or so of dehydration.

You will also find that air drying times for all the foods mentioned in this chapter will be reduced by 3 hours on the average; so don't set the oven and then forget about it until 10 or 12 hours have elapsed. You need to remain in fairly constant attendance.

DEHYDRATING

Probably the best type of dryer is the dehydrator that is designed and built specifically to reduce moisture in foods. These commercial dehydrators are well worth the cost, since they don't depend on sunny weather and they don't tie up your oven all day.

To prepare food for dehydrating, use the same methods you used for air drying. The only differences will be that you can't dry as much food at a time as you can in an air or solar dryer.

Since the drying time is less, in the same amount of time you can dehydrate about the same amount of food.

Better, however, is the fact that you don't need to worry about rain and insects. You can spread the food on the racks in the drawers, set the controls, and leave the food dehydrating even overnight if you wish. About the only care you will need to give is the occasional rotating of drawers so that one rack of food will not always be the closest to the source of heat.

Anything that dries well in the oven or air dryer will dry superbly in the dehydrator. You may need to turn large slices of food once every 2 hours, or perhaps less often, and at times you may need to crack the drawers slightly (about an inch or two) in case the temperature inside the dehydrator rises too far. One of the tricks to drying food is to extract the moisture slowly; you don't want to cook the food (Figs. 6-8 through 6-12).

You will enjoy dehydrating the food almost as much as you will enjoy eating the finished product. You need to use no electrical power to keep the food once it is dehydrated, and you need only a fraction of the shelf space you need for canned or frozen foods.

FREEZING

Because freezing is such a simple and effective way to preserve foods, millions of people own freezers. But many of these people may not realize the amazing variety of foods that can be frozen beautifully if a few basic methods and precautions are observed.

The major complaint about frozen fruit is that it tends to turn dark and lose flavor when frozen. You can eliminate both of these problems very easily, however, if you will use ascorbic acid or a brine solution to dip the fruit in before you freeze it.

Ascorbic acid is not a food additive to be feared. It is, in fact, a very desirable additive, since it is only Vitamin C. By using ascorbic acid you not only prevent the discoloration, but you also preserve flavor and add nutrients to your food.

You can buy packets of absorbic acid at most stores where canning supplies are sold. The powdered form works very well, so buy it rather than tablets, if possible.

Dissolve the proper amounts into a small amount of water and use the acid either in syrup packs, sugar packs, unsweetened packs, or containers of fruit juices.

If you make a syrup, you can sprinkle the

Fig. 6-8. To dry pumpkin, cut the pumpkin in half, scoop out the seeds, and cut the melon into oval sections.

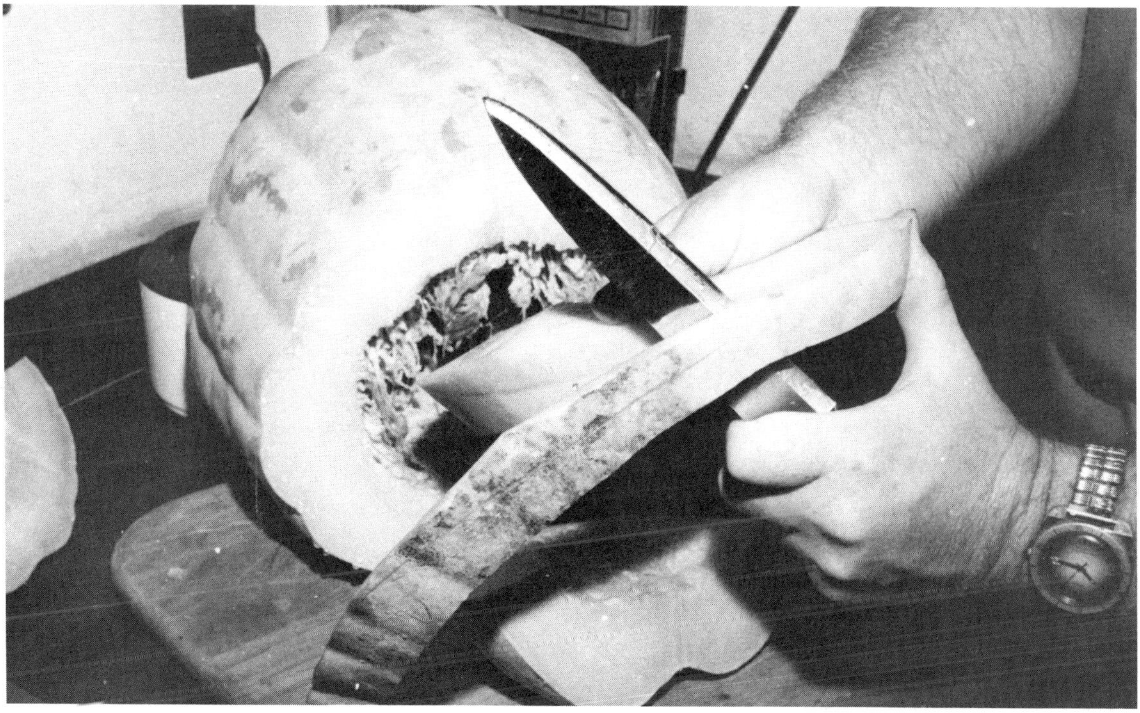

Fig. 6-9. Peel the melon, making certain to remove all rind.

Fig. 6-10. Cut the sections into cubes and arrange on a tray.

Fig. 6-11. Put the rack in the dehydrator, preferably in the top drawer. The pumpkin is so juicy that if it is too close to the heat it will dehydrate too fast.

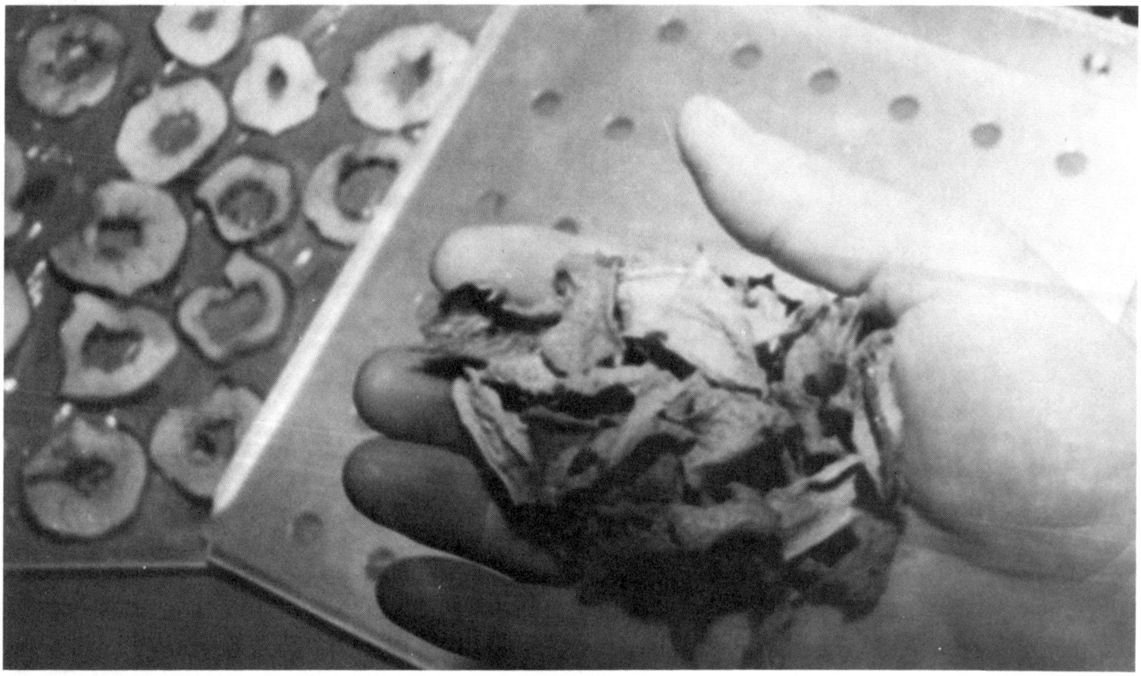

Fig. 6-12. Dried pumpkin is only a tiny fraction of the original volume. A rackful can be held in the palm of one hand.

proper amounts of ascorbic acid into the syrup and then pack the fruit according to instructions. In sugar packs you can sprinkle the acid over the fruit and then add the required amounts of sugar.

If you do not want to add sugar to your fruits, you can sprinkle the ascorbic acid over the fruit just before you pack it to freeze.

When freezing fruit juices, you can add the acid to the juices, stir well, and freeze as instructed.

If you prefer not to use the ascorbic acid, you can make a brine mixture by dissolving two tablespoons of salt into a gallon of water. Before you pack the fruit you can soak the slices in the solution for 5 to 10 minutes, but do not exceed 20 minutes, and then pack and freeze.

Virtually all types of fruits can be frozen, and we will suggest methods for major categories of fruits usually frozen with good results.

Apples

For apples, choose the best possible fruit to begin with. Apples should be crisp and firm, never mealy or soft. After you wash and peel the apples, core them and slice into sections suitable for use in pies. You can use a light to heavy syrup, depending upon your tastes and needs to pack the apple slices in, or you can mix 1/2 cup of sugar with a quart of sliced apples, coating all slices thinly with sugar. Pack and freeze.

For applesauce, cook the apples until they are soft and tender. Cook slices rather than halves for faster cooking, then puree and add sugar to taste, usually 1/4 to 1/3 cup per quart of cooked apples.

While you are cooking apple slices you will cook a great deal of juice from the apples. You may save this juice until you have a sufficient quantity to freeze or you may use it in making apple jelly. To freeze the juice, simply pour it into a freezing container and place the container into the freezer.

Other Fruits

Apricots should be halved and pitted, then dipped into boiling water for about 30 seconds. Transfer the halves to a container of cool water, then drain the cooled sections. You can pack the sections into your container and put a syrup over the apricots, or you can mix the ascorbic acid solution and treat the apricots by sprinkling the mixture over the halves and then mixing the halves already treated with sugar. You will need 1/2 a cup of sugar for each quart of apricot halves you plan to freeze. Pack in juice or dry pack and freeze.

One of the best methods for freezing blackberries, similar berries, and other closely related fruits is to wash the berries and then pack them into containers. Pour a syrup over them and freeze. Always remember to leave some space at the top of the container to permit expansion as the fruit and syrup freeze. Another effective and quick way is to sprinkle 1/4 of a cup of sugar over the berries, pack, and freeze.

Cherries should be stemmed and pitted, washed carefully, and drained. You can shake the cherries in a bag containing 1/2 cup of sugar for one quart of sour cherries or 1/4 cup for sweet cherries, then pack and freeze. You can also pack the cherries in a syrup mixture and freeze.

Currants and cranberries can be washed and drained, then packed in syrup and frozen, or they can be frozen after being sprinkled with sugar and packed.

A good way to freeze figs is to wash and peel them, then shake the figs (a good plan is to halve them lengthwise) in a bag containing 1/2 cup of sugar for each quart of fig halves. Pack and freeze. You may also pack fig halves into a freezer carton and then pour a medium syrup over them and freeze.

You can freeze oranges, grapefruit, and pineapples by packing them in a syrup made of juice, water, and sugar. You can also squeeze juice from fruit and freeze the juice unsweetened, or you can, if you prefer, add 2 or 3 tablespoons of sugar per quart of juice.

If you have a good supply of grapes, you can wash and stem them, and seed if necessary. Pack in freezer cartons and pour a medium syrup over them. If you prefer to squeeze the grapes for juice, you can freeze the juice as it comes from the grapes.

Melons should be frozen in small amounts—2-inch cubes, for instance. After packing, pour a light syrup over the fruit and freeze.

You can freeze peaches in syrup or sugar. Use good, firm peaches, peel, remove seeds, and either freeze halves or slices. Pack into cartons or, if you prefer, sprinkle ascorbic acid over halves and slices. After you pack the peaches, pour a syrup over the sections and freeze. You can also shake a quart of halves or slices in a bag or other container into which you have poured 1/2 cup of sugar, and then pack and freeze the coated peaches. Pears can be handled in the same way.

Plums should be washed, halved, and pitted, then packed in a syrup to which 1/2 teaspoon of ascorbic acid has been added. If you wish, you can omit the acid. You can also sprinkle sugar over the halves or shake them in a bag containing 1/3 cup of sugar for each quart of plums.

If you enjoy rhubarb pie and have access to a supply of rhubarb, wash stalks well and cut off ends and leaves. Shake it in a bag containing 1/2 cup of sugar and one quart chopped rhubarb stalks, or pack one inch segments into containers and pour a syrup over the contents. You can also dip the stalks into boiling water for 1/2 to 1 minute, cool, and pack in syrup.

Strawberries can be washed, then capped, and shaken in a bag containing 1/4 cup sugar and one quart of berries. You can pack the bag with berries, then dump the sugar over the berries and shake the bag so that the sugar will work down to the berries nearer the bottom as well as cling to those on the top. You can also pack in a thin syrup if you prefer.

Tomatoes

Tomatoes can be frozen easily and well, but they can be used only for cooked dishes after thawing. To date no one has found a way to freeze tomatoes that can be thawed, sliced, and eaten in salads or sandwiches the way fresh tomatoes can.

To freeze tomatoes, choose ripe, firm, and sound tomatoes that are free from blemishes. Dip them for 1/2 minute into boiling water and the peel or skin will slip off easily. Then halve or quarter the tomatoes, pack into containers, and freeze. You can also cook the tomatoes for five to eight minutes, then pour the sections into containers and cover with juice. If you have juice left over, you can freeze it, too, and later use it for cooking or drinking.

Virtually any type of fruits discussed here can be frozen raw and dry if you prefer. You can sprinkle a small amount of sugar—1/4 cup generally—and freeze without blanching or cooking. There will be some slight discoloration if the fruit is exposed to air for more than a few minutes.

ROOT CELLAR STORAGE

It is a rare home that has a root cellar in it today, but, even if you don't have a cellar, there are certain foods you can store in a similar fashion. If you have a basement that is dry, you can in effect practice root-cellar storage and preservation of food.

In terms of economy, this method is by far the best, since there is no power to operate a dehydrator, to cook the food for canning, to power the pressure canner, or to operate the freezer. In addition, there is very little special preparation of foods necessary.

What, then, keeps everyone from using root-cellar storage? For one thing, the previously mentioned shortage of cellars. For another, there is a limit to how long the food can be kept before decay, insects, or other physical problems attack and ruin it.

A ***root cellar*** is, by definition, a storage area with a dirt floor that is three or four feet below the surface of the ground outside-more if possible and practical. To make one you can dig a basement, or the beginning of one, and when you have reached a suitable depth, you can build a small outbuilding over the excavation. If possible dig and pour concrete footings to keep rodents from digging tunnels into your cellar. It is amazing to see how much food can be eaten, or at least ruined, by one squirrel in one week. Rabbits can devour bushels of food in a remarkably short time.

Since insulation is crucial to root cellar storage, you must have very thick walls and preferably a double-wall structure that permits a dead-air space. If you are handy with tools and have a supply of cement blocks, you can build the outbuilding with blocks and fill in the holes with sawdust as you complete each course of blocks. Build another wall inside the first one, if your time, energy, and block supply permits. You will need a roof and air vents for the building, too, since you can't afford rain leakage, humidity, or condensation retention inside the storage area.

If building a root cellar is too much trouble for the limited use you would get from it, you might investigate the crawl space under your house to see if you could erect an inside wall structure that would give you a small room in which to store your food.

You could also try your basement. You will have a concrete floor, but you can still store a large amount of food if the space is cool and dry.

Some fruits simply will not keep in root-storage conditions. Peaches will develop soft spots, and the rapid enzyme action will reduce a bushel of peaches to a mass of decay in a matter of a few days. Pears keep reasonably well under these conditions, but plums, apricots, figs, cherries, and other fruits last only a very short time.

Tomatoes

To store tomatoes in this fashion you need to pick them while they are very hard and still green. Ripe tomatoes will not last, so don't even try to preserve them in this fashion.

It is a very good idea to plant late tomatoes with the idea that the fruit will never have a chance to ripen outdoors. Pick the tomatoes while they are in the lighter shades of green, and wrap each tomato individually in half a sheet of ordinary newspaper. As you wrap each fruit, place it in a box or basket and be sure the paper does not come unwrapped. Transfer the box to the basement or root cellar and see that it stays closed.

If you have no other way to store tomatoes, you can push the box under a bed or store it on the floor of a closet. Check the tomatoes every two weeks or so by unwrapping one of the top fruits and examining it for signs of decay or ripeness.

Although such tomatoes will not keep nearly as long as frozen, dried, or canned tomatoes, you can have fresh, red tomatoes until January, perhaps later. For all intents and purposes they are as good as tomatoes picked fresh from the garden.

Apples and Pears

You can store apples in a root cellar or similar area or space in the autumn and keep them until the following spring unless they get too warm so that enzyme action can start, or get wet, or freeze. You can wrap them the way you did tomatoes and they will keep well. Even if the peeling shrivels a little, do not worry. Water is being dehydrated by nature's own process and what is left is a greater percent of sugar than when the apple was firm and plump. In a very real sense the apple has a better taste that it would have had when it was picked off the tree.

If you prefer, you can wad sheets of newspaper and fill the bottom of a box with the soft balls, and then you can start to fill in with apples. Add another layer of newspaper after the apple layer is filled. Follow this plan all the way to the top of the box, but be sure to pack wadded newspapers on all sides and save some room at the top for a layer of papers as added insulation.

Pears can be stored in a similar fashion. You can pick both apples and pears before they are ripe and they will continue to ripen in the containers. Many fruits and berries, such as strawberries and plums, will not ripen after leaving the parent tree, and others, like bananas, will ripen slowly in a warm room and much slower in a cool or refrigerated area.

ESTIMATING YIELDS

As a last thought for this chapter, it might be very helpful for you to know about how much food to expect from a bushel of produce. How many quart jars will you need to have on hand if you have a bushel of apples? How many freezing bags will you need for a bushel of tomatoes?

The following information is based on average productions, and of course there might be a great variation in the moisture content of some fruits or tomatoes. Many variables must be taken into consideration, but it is still possible to give you a general idea how much yield you can expect.

A bushel of tomatoes, which will weigh 50 to 60 pounds, will produce about 18 to 20 quarts of canned or frozen tomatoes.

A bushel of apples, weighing 40 to 50 pounds, will yield about 25 quarts of preserved food.

One bushel of peaches will produce between 20 and 25 quarts from the original 40 to 50 pounds.

Pears, which weight slightly heavier than apples and peaches—up to 60 pounds per bushel—will make about 25 quarts when canned or frozen.

A bushel of plums will produce nearly 30 quarts from the original 50 pounds.

Fifty pounds of apricots will make 30 quarts of canned or frozen food, and a similar amount of blackberries or dewberries will yield the same number of quarts.

It will take about 3 pounds of grapes to make 1 quart, and 60 pounds of strawberries will make about 27 quarts of frozen or canned fruit.

7

Preserving Low-Acid Vegetables

Low-acid and neutral vegetables are those garden foods which are so low or lacking in natural acids that they are very mild in tartness and sometimes rather bland in flavor. Low-acid or neutral vegetables include corn, squash, beans of all kinds, okra, beets, potatoes, pumpkins, asparagus, and many others. In fact, nearly all vegetables are considered low acid.

All these vegetables must be given more steam pressure for longer periods of time, since there is no acid to help kill and prevent recurrence of bacteria that can spoil food and cause serious food poisoning.

Since there are no acids to help preserve food, you must never under any circumstances can these vegetables by the open-kettle or water-bath methods. There is no way you can get the temperatures high enough to kill all of the harmful micro-organisms.

We know that many old-timers in home food preservation insist that they have been canning vegetables by the water-bath method for decades and the worst they have experienced is an occasional jar of food that went bad and had to be discarded. In truth, we, too, have canned by the water-bath method, and we have eaten and enjoyed the green beans, new potatoes, corn, and all the other foods we are asking you not to can in a water bath.

There was a time, not very long ago in terms of American history, that the water-bath canner was really the only kind we knew about. The pressure canner as we know it had not been invented, and home freezers were, if on the market at all, very rare. So rural people in particular had to can or depend upon drying and grocery stores to help them survive during the winter months.

Today there is no reason to depend upon the water-bath canner except for fruits and tomatoes. Use the pressure canner and freezer instead, or dehydrate or store in other ways all the low-acid vegetables you preserve.

These low-acid foods are very easy to work with, and they look great on shelves and are delicious to eat; some, in fact, seem to taste better after canning and freezing than they did while fresh. Many people, for instance, enjoy canned green beans better than fresh ones.

PREPARATION

Many of the vegetables in this section can be canned, frozen, or dehydrated if you chop them into small pieces or slice them. If they are small, you can halve, quarter, or leave them whole.

You will want to dice squash, for instance, or, if the squash is small, you can halve it. Corn can be preserved by cutting the kernels off the cob, or you can preserve the entire ear as corn on the cob. String beans can be put up whole or broken into 1-inch or slightly larger sections. Green peas can be shelled and then preserved. Asparagus can be preserved whole, as can greens of all types. Greens, too, can be chopped before processing.

In short, the work is very easy, and the preparation is so varied that you have a great deal of latitude in your approach.

As with nearly all steps in food preparation, you should wash the vegetables thoroughly and use a brush to clean the tiny crevices you find in many vegetables. If the food is in a hull or shucks, remove the outer covering. You will still need to check the food carefully despite the fact that dirt could not get inside. You may find insect damage or spots of decay in beans, and corn ears have silks that must be removed. Corn is also invariably attacked by a variety of insect pests. Certain worms thrive in corn and cause extensive damage in a very short time.

After washing the vegetables, you will need either to blanch or to cook, in many instances. We recommend blanching or cooking to the last minute of the prescribed time, since failure to cook long enough can be a failure to kill the micro-organisms that are so damaging.

Specific preparation information will be provided for particular foods in the discussion of how to preserve the various vegetables discussed in this chapter.

DANGER OF FOOD POISONING

While the danger of serious illness or death by food poisoning is a grim reality, the possibility can be virtually eliminated if you follow the instructions for your processing.

The one poison that is mentioned repeatedly in this book is botulism, which probably causes more deaths from food poisoning than any other type of poison. The major culprit is home-canned green beans, followed by corn, spinach, beets, and chard. Some cases have been traced to improperly processed meats. String beans apparently cause the highest number of cases because of the vast amounts of beans canned in this country.

Symptoms of botulism are apparent from 2 hours up to 36 hours following ingestion of the contaminated food. Common symptoms are double vision, difficulty in speech, difficulty in swallowing, and very heavy or labored breathing. Death results usually from paralysis of the respiratory system.

It has been demonstrated that if you use the pressure canner for low-acid foods, the botulism spores will be destroyed very effectively, and the food will be completely safe, if the jars are properly sealed.

Remember that any time you open a jar of food and there is a foul odor, discard the food. Don't even put it outside where family pets may eat it. Also, if a can of food is swollen, discard the contents immediately. Once in a while you may start to open a jar and you will notice that the lid is puffed upward. This food should not be eaten. In fact, in most instances in which the lid is puffed outward, the lid will be forced off the jar in a few more hours.

We also recommend that when you open a jar, check to be sure that the lid is still curved downward and that the jar is sealed well. When you pry the lid off you can usually hear the suction of air into the vacuum created by the canning process. If there is no audible suction and if the lid came off very easily, you should be wary of the food in the jar.

With green beans we recommend that, even after you have been assured that the seal was good and that there is no real likelihood of botulism, you boil the beans for 20 minutes before you eat them.

PRESSURE CANNING

Your first step in pressure canning is to select proper

vegetables to can. The selection should be based on a thorough visual examination of the food in question. Remember that any food you would not wish to prepare for your next meal should not be used for canning unless you can acquire no other. Choose vegetables that are tender and just slightly immature if possible. Be sure that there are no insect-damaged spots and that there are no bruised areas on food before you put it into the jars. If insects have damaged the food, cut out the offending spot.

Hot-Pack Method

You should have an ample supply of sterilized jars, lids, and rings. Wash all foods carefully and prepare them for precanning cooking. It is a good idea, unless otherwise noted, to use the cooking water as a liquid to cover the food in the jars, because during cooking the vegetables will lose some of their nutrients, and these nutrients will remain in the cooking water. If you pour this water over the vegetables, you save the valuable nutrients.

Asparagus. Asparagus should be picked when young and tender. Cut away all tough ends and then wash the whole stalk carefully. Use a food brush to get into areas where dirt may be difficult to spot or remove. Cut the asparagus into one inch or slightly longer pieces and drop into boiling water to cook for three minutes. Remove from the water and pack immediately into hot, clean jars and cover the asparagus with boiling water. You may add a scant teaspoon of salt if you wish; if you must restrict your salt intake, omit the salt and put on lids and rings. Tighten the rings and put the jars into the pressure canner and follow the directions given in Chapter 6 for processing foods in the pressure canner.

Process the asparagus, if you use quart jars, for 40 minutes; if you use pint jars, 25 minutes of pressure time will be sufficient. Pressure should be kept at 10 pounds at all times.

Green Beans. Select young and tender green beans, wash thoroughly, and break off ends and string the beans, if necessary. When you snap the ends off, there will be a thin fiber, that runs down the length of the bean on both sides, which will be freed. Pull this fiber or string the rest of the way off and discard. You may cut or break the beans into one to two inch lengths or you may leave them whole.

Put the bans into a pot, cover with water, and bring to a boil. Cook for 5 minutes and then pack the beans into hot, sterilized jars. Adjust lids and rings, tighten the rings by hand, and put the jars into the pressure canner. Process the beans at 10 pounds of pressure for 20 minutes for pint jars and 25 minutes for quarts.

Lima Beans. For lima beans, shell the beans and wash, then drop the beans into boiling water. The water will stop boiling when the beans are added, but leave the heat on until the water returns to a boil. Leave the beans in the boiling water for no more than 1 or 2 minutes and then pack into hot, clean jars. Hand-tighten the lids and rings and place jars into the canner. Process at 10 pounds of pressure for 35 minutes for pints and 50 to 60 minutes for quarts. Be sure to cover the beans with boiling water before you put on the lids.

Beets. Beets can be packed whole if they are small, but, if they are large, you can quarter or slice them. To prepare beets for canning, wash carefully and cut off the stems about one inch from the beet top. Leave the inch-long stem on in order to prevent the beets from 'bleeding' and losing color. Drop the beets into boiling water, cook for about 15 to 20 minutes, then slip the skins off, cut away the stems and roots, and pack into hot sterilized jars. Cover with boiling water after you add one teaspoon of salt, if desired, and close jars and tighten lids.

Process at 10 pounds of pressure. Pints should be processed for 25 minutes; quarts should be processed for 45 to 50 minutes.

You can save the beet tops or greens and cook these and eat them just as you would prepare other garden greens. Check with your nursery or seed store to determine which beet greens are best for eating. You can also make delicious pickles from beets, and the process will be described in Chapter 8.

Broccoli. To can broccoli, remove the outside leaves, wash thoroughly, and cook for three to four minutes. Pack into hot, sterilized jars. Tighten lids after you cover the broccoli with boiling water. You may add a teaspoon of salt if you like. Be sure to use fresh boiling water rather than the water the broccoli was cooked in. Process at 10 pounds of pressure for 35 minutes for pints and 40 minutes for quarts.

Brussels Sprouts and Cabbage. Brussels sprouts and cabbages must be cooked for 3 minutes. Pack the hot vegetables in jars, but do not use the cooking liquid to cover. Use fresh boiling water instead. Process pints at 10 pounds of pressure for 35 minutes or quarts at 10 pounds of pressure for 40 minutes.

Carrots. Carrots should be young and tender. While washing carrots, scrape them and sort according to size. Carrots may be canned whole, sliced, or cubed. Place prepared pieces in a cooking vessel and cover carrots with boiling water. Bring entire contents to a boil. Pack hot in jars adding one teaspoon of salt to each quart, if desired. Cover the contents with the boiling liquid in which the carrots were cooked. Process pints in the pressure canner at 10 pounds of pressure for 25 minutes; quarts at ten pounds of pressure for 30 minutes.

Corn. Corn should be canned that same day it is picked. Some food experts warn that corn left standing for just one night should not be used for canning. For best results use only fresh and tender ears. There are two popular methods of preparation for corn: cream-style and whole kernel.

For cream-style corn cut kernels from the cob and then scrape the cob to remove all the juice. Do not allow any part of the cob to be included. To each quart measure of corn and juice add one pint boiling water. Bring entire contents to a boil. Pack hot, adding 1/2 teaspoon salt to pints if desired. Process in a pressure canner at 10 pounds of pressure for 85 minutes for pints and quarts. Pints of cream-style corn provide generous servings for a family of four; therefore, pint containers may be used almost exclusively.

Whole kernel corn is prepared by cutting the corn from the cobs, but no scraping of cobs is done. For each quart of kernels add one pint of boiling water and 1/2 teaspoon salt, if desired. Pack hot kernels loosely in jars and cover with the boiling liquid. Process in a pressure canner at 10 pounds of pressure for 55 minutes for pints and 85 minutes for quarts.

Canning whole ears of corn is not recommended under any circumstances, since the density of the cob does not allow a high enough steam pressure to be safely reached.

Greens. Freshly picked greens of all kinds may be successfully canned. In preparation, the thorough washing of the leaves to remove any grit and the discarding of any yellowed, tough stems or roots are highly important. Once leaves are sorted and cleaned, place them in a steamer or bundle them in a cheesecloth bag and steam greens about 10 minutes or until wilted.

Pack hot into jars adding 1 teaspoon salt to each quart, if desired. Cover the contents with boiling water or with the liquid used with the steaming if the leaves did not cause it to become bitter. Process jars in a pressure canner at 10 pounds of pressure for 45 minutes for pints or for 70 minutes for quarts.

Mushrooms. Mushrooms are easily canned since the basic steps are simple: wash, slice, cook 3 minutes, pack hot. A few suggestions will make the final product more enjoyable, and these extra steps do not take extra time. When preparing mushrooms, trim away any discolored parts or tough stems. To ensure that all soil is removed soak mushrooms in cold water for about 10 minutes. Then wash again in clean water. Steam pieces or small, whole mushrooms for 3 minutes, or cook slowly in a covered saucepan with no liquid added for about 15 minutes. Pack hot in either half-pint or pint jars since small amounts of mushrooms are usually required in most recipes. To each pint add 1/2 teaspoon salt and 1/8 teaspoon ascorbic acid—for better color—if desired. Cover contents with boiling water and process in a pressure canner for 30 minutes for both pints and half-pints at 10 pounds of pressure.

Despite the fact that mushrooms can be canned, keep in mind that some mushrooms are deadly poisonous and can and will cause sudden death. ***Toadstool*** is really a German term (todestuhl) meaning chair or stool of death.

If you don't know your mushrooms, don't start picking and canning. Some mushrooms are poisonous until cooked; others are non-toxic until cooked. Buy your mushrooms from a reputable and reliable source unless you are really an expert. Even if you are an expert, exert the greatest caution.

Okra. Okra may be canned in whole pods or sliced. Small, tender pods work best if the pods are to be used whole. Wash okra thoroughly before slicing. Place pieces or pods in a cooking vessel and cover with boiling water. Bring entire contents to a boil and continue cooking for 2 minutes. Pack hot in jars adding 1 teaspoon salt to each quart, if desired. Cover okra with boiling water and process in a pressure canner at 10 pounds of pressure for 25 minutes for pints and 40 minutes for quarts.

A tasty combination of vegetables is okra and tomatoes. Again use only tender pods of okra and sound tomatoes. Bring sliced pods and tomato sections to a boil. Remove from heat and pack hot, adding 1 teaspoon salt to each quart if desired. Processing times are the same as for okra.

Green Peas. To can green peas, two washings are necessary. First, wash the freshly picked pods. Then shell the peas and wash them. Place peas in a cooking vessel and cover with boiling water. Bring entire contents to a boil. Remove from heat and pack peas loosely into jars adding 1 teaspoon salt to each quart if desired. Cover peas with the boiling liquid from the cooking of peas. Process the jars in a pressure canner at ten pounds of pressure for 40 minutes for pints and quarts. A pint of green beans provides a generous serving for a family of four; therefore, you may want to use only pint jars for canning this vegetable.

Peppers. Peppers, while they may be successfully canned, are much easier to freeze; so we recommend freezing entirely.

Potatoes. Small, whole potatoes are canned in one of two ways. New potatoes that are no more than 2 inches across may be washed and scraped. Cover the potatoes with boiling water and bring entire contents to a boil. Continue boiling for 10 minutes. Pack hot, adding a teaspoon salt to each quart if desired. Cover contents with boiling water and process in pressure canner at ten pounds of pressure for 40 minutes for pints and quarts.

Another method of canning small potatoes is to wash the potatoes, cook them for five minutes, then slip the skins, and pack potatoes hot into jars. Cover contents with boiling water and process at times and pressure given above.

Pumpkin. Pumpkin is easy to can but decidedly deceptive in the amount of the yield. When the pumpkin is cooked down and the moisture extracted, you have much less than you had anticipated. But it is still a nice food to have around, since it makes a delicious pie or custard.

To can it, peel the pumpkin—a much harder job than you might think—and then chop it into 2-inch cubes. Cover with water in a large pot and bring to a boil. Pack hot and loosely in sterilized jars that are also piping hot. Put on lids and rings. Hand-tighten and place jars in pressure canner. Process at 10 pounds of pressure. Pint jars need to process for 55 minutes and quarts run for 95 minutes.

Summer Squash. You can process summer squash by selecting small and firm squash, washing them thoroughly, and cutting off the stem and blossom ends. Now slit in half lengthwise. If the seeds are small, you can leave them in; if the seeds have developed, scrape the seed cavity clean with a spoon. You can cook the squash halves intact or you can cut them into cubes. If you use very large squash, cut into cubes. Cook the squash for 1 minute, or slightly longer, in boiling water, then pack into hot, sterilized jars. Cover the squash with the cooking water. Close by placing lids on jars and adding rings. Hand-tighten and process at 10 pounds of pressure for 30 minutes for pints and 45 minutes for quarts if you canned halves and 40 minutes if you cut the squash into cubes.

Winter Squash. Winter squash can be canned in much the same way as summer squash. Cook until the boiling point is reached, then pack in hot, sterilized jars. Cover the squash with the same water in which it was cooked. Close jars and process at 10 pounds of pressure. Maintain pressure for 1 hour for pints and 1 1/2 hours for quarts.

Succotash. You can prepare a succotash mixture by cooking kernels of corn and lima beans for 5 minutes each but in separate containers. Combine the two vegetables and cover with the water in which they were cooked. Process pints for 65 minutes and quarts for 85 minutes at 10 pounds of pressure.

Sweet Potatoes. Sweet potatoes are excellent when canned. You can wash and drop potatoes into hot water, boil until the skins are easily removed, then cut into cubes and pack in hot, sterilized jars. You can instead peel first, cube, and cook for five minutes and pack into hot jars. Cover with the water in which they were cooked. After you pack the potatoes into the jars, you may want to use a spoon or knife to work out air bubbles. Then process pints for 1 hour and quarts for 1 1/2 hours at 10 pounds of pressure.

Soup. You can make a soup mixture by cutting any of several vegetables into small cubes. In the case of beans or corn kernels, leave them whole. Use corn, peas, celery, chopped string beans, lima beans, carrots, tomatoes, onions, and other vegetables that might blend with this mixture. Heat to the boiling point and pack in sterilized, hot jars. Pour the cooking liquid over the mixture and close the jars. Process at 10 pounds of pressure. Run the jars as long as you would run the vegetable with the longest processing time.

Other Vegetables. These are not the only vegetables that can be pressure canned; however, this listing gives a representative sampling of how others may be processed. The same general principles apply to all vegetables, with few exceptions, if any.

Cold-Pack Method

Thus far we have dealt with the hot-pack methods of pressure canning. But several vegetables, including asparagus, lima beans, green or string beans, pole beans, carrots, corn cut off the cob, peas, and squash can be cold packed.

To cold pack, wash vegetables thoroughly, break or cut into desired sizes, pack into jars, and pour boiling water over the food. Process, after tightening lids, by the following times, all at 10 pounds of pressure. The time given are for pints and quarts, respectively: asparagus, 25 and 40; lima beans, 40 and 50; string beans and pole beans, 20 and 25; carrots, 25 and 30; corn, 55 and 85; peas, 40 and 40; summer squash, 25 and 30.

Store the vegetables as you would store fruit and tomatoes. Keep them in a cool place but one in which they are protected against freezing.

DEHYDRATING

Dehydrating low-acid vegetables is remarkably easy. You will need to blanch and pretreat them. You should store the dried food in small amounts rather than in a huge can or other container, because if one piece still has too much moisture, enzyme action can start or decay will eventually destroy the whole package. So plan to store in quart or smaller sizes of containers.

Preparation and Drying Time

The basic principles are much the same for all vegetables, but the preparation varies. Again, the drying time will depend on the type of food, the amount of moisture, the heat inside the dehydrator, and the proximity of the food to the source of heat.

If the food is stiff, leathery, and tough, it is probably ready. Sample a piece, cut it to see if any moisture seeps out or can be squeezed out. If moisture drops appear, leave the food in the dehydrator another hour, then check again.

Asparagus. To dry asparagus, gather the sprouts or shoots as soon as they are ready. You want young and tender growths, and you should cut these down the center lengthwise and then chop the sections into one-inch lengths. You should blanch by dipping into boiling water for 3 or 4 minutes. Drain, blot dry, and spread on the dehydrating racks or trays. Check the trays every hour and turn the food at every 1/2 hour or hour. When the segments are dry and brittle, remove them, pack into the air-tight containers, and store.

Dried Beans. It is difficult to find any food easier to prepare than dried beans. All you really need to do is leave the beans on the plant until the pods start to turn dark, then pick the beans and spread the pods in a sunny and dry place and let them continue drying for another two or three days. When you shell the beans, spread the beans again to dry. At this point you can dry them in a dehydrator until they are very hard and brittle. Store in insect-proof, air-tight containers. This process works for pinto beans, lima beans, black-eyed peas, crowder peas, and any and all other forms of dried beans that are commonly grown.

Beets. Prepare beets for drying just as you prepared them for pressure canning. Cut stems off one inch from the top of the beets and drop the beets into boiling water and cook until the skins slip off easily. Remove from water, cut off stems and roots, and slice beets 1/4 inch thick and blot dry. Spread the beets on the drying racks and let them dehydrate for about ten hours or until you cannot squeeze moisture from a fresh cut and no droplets of moisture seep from the incision.

Broccoli. Select stalks of broccoli that are fresh and tender. Split the stalks and chop the sections into 1-inch lengths. Plunge the broccoli into boiling water and leave for 3 or 4 minutes, then drain well. Spread on dryer racks and leave until it is crisp but not brittle. Turn every hour or so and remove from the dryer and store as soon as it reaches the proper state of dehydration—you may have to leave it overnight since it dries so slowly—usually after 10 to 15 hours.

Brussels Sprouts. Brussels sprouts should be picked when still young and tender, and you should, after a thorough washing of the sprouts, split them in half and drop them into boiling water. Leave them for five minutes, remove from water and drain. Spread the halves on the racks and allow them to dry from 6 to 8 hours on one side, then turn and dry the other side. Pinch or bite one of the halves. If it is tough and brittle, it is dry enough to store.

Carrots. To dry carrots, select young, crisp carrots and slice them into sections 1/4 inch thick. Drop slices into boiling water and leave for 3 or 4 minutes, then drain. Dehydrate for 10 hours or until the slices are leathery but pliable.

Cabbage. You can dry cabbage by shredding leaves as if you were planning to make salad. Dip the leaves into boiling water and leave it for two minutes, then drain and spread on racks. The cabbage will turn dark and become very crisp in about 8 hours. Move the shredded leaves around so that there is an equal exposure to the heat, and wait until all the leaves are crisp and light brown or the color of dried grass.

Cauliflower. Cauliflower should be chopped into tiny sections and blanched in boiling water for 2 minutes, drained, and spread on the dryer racks. Allow it to dehydrate for about 8 or 9 hours at least. Spread and turn the sections every 2 or 3 hours. You can dry larger sections but the time is expanded considerably—up to 15 or even 20 hours, depending on the thickness of the sections.

Corn. To dry corn, you may choose one of several methods. The simplest way is to leave it on the cob and the cob on the stalk and allow nature to dry it. You can then gather the corn, remove shucks, and store in a safe place or, to be sure moisture has been removed, you can spread the shucked ears in a sunny place to dry further. You can shell the kernels off the ear easily when it is dry enough to store, or you can actually store the cobs and corn together.

A second way is to gather the corn, pull the shucks back so that the corn is exposed to sun and air, and suspend in a dry area.

But any corn left on the cob and in the field will be susceptible to attacks by insects and birds, so the best way is to gather the corn while it is still tender and juicy, cut the kernels from the cobs, and blanch the kernels in boiling water for about three minutes. Spread to drain and transfer to the dehydrator rack, where it should be allowed to stay for 8 or 10 hours. Spread and move the kernels around for even exposure to the heat.

You can also dry corn on the cob by dipping the corn into a pot of boiling water, leaving it for 2 or 3 minutes, and then putting the entire ear into the dryer or dehydrator. Allow 15 hours to dry (Figs. 7-1 and 7-2).

Greens. For an interesting experience, try drying greens of all types: turnip greens, mustard greens, collards, spinach, kale, rape, chard, and certain beet greens (ask your garden supplier which are best). Pick and wash the greens and leave them whole, except for collards, which often are so large that they cannot be handled well.

Steam blanch for about 3 or 4 minutes, then

Fig. 7-1. Well-developed ears of corn will have kernels of large, uniform size all around the ear. As you press your thumbnail into a kernel, if the corn is fresh, a milky juice will appear.

drain. After draining, the leaves should be spread in several layers (if your space is limited) on the racks and then turned often as the leaves start to shrink. Drying time is only about 6 to 8 hours. Handle carefully when you start to store the greens since the brittle leaves—almost the same texture as dead leaves from a tree—will crumble easily.

Cucumbers. One food that is fine dried is the cucumber. Pick cucumbers that are small enough that the seeds have not grown to their maximum size. Wash and slice the cukes into thicknesses of about 1/4 inch. Do not blanch. Spread the slices on the racks and dry until nice and crisp—crisp enough to be eaten almost like potato chips.

Lettuce. You can dry lettuce in the same way that you dry cabbage. Don't use the tough outer leaves if you use head lettuce, and don't use the stalk. The rest of the process is the same as cabbage or greens.

Mushrooms. To dry mushrooms, first buy mushrooms that you know to be safe and then cut the bottom of the stem, where it has been affected by exposure to the air, and wash quickly and well in cool water. Then spread them on the dryer racks and leave for about 15 hours. If you chop or slice into thin segments, they will dry much faster.

Okra. Okra should be picked while young and tender and then washed and sliced the same thickness you would want if you planned to fry the slices. Blanch in boiling water for 2 minutes and spread, after draining, on the dryer racks. Leave until good and crisp.

Onions. One of the easiest ways to dry onions is to peel off the outer layers of the plant and then slice the onion the same thickness you would use for a sandwich. Separate the rings and spread on the rack. They will dry to a fine crispness in about 7 hours in a dehydrator.

Peas. Peas should be picked when young and slightly immature. Shell the peas from the pod and then blanch in boiling water for 2 minutes. Drain and spread on the racks to dry. In about 12 hours or slightly more you will notice that the peas have shriveled and have been reduced to about 1/2 their former size. They should be as hard and brittle as a peanut when you remove them from the dryer.

Peas such as crowder peas can be dried like pinto beans and other shelled beans. Again, they dry best on the vine, but gather them before they are completely dry, since the pod, which aids in propagation, will snap open with a very audible popping noise and shoot the peas for several feet in every direction and you will lose more than you gather.

Peppers. Peppers dry beautifully if you wash, split the pod open, and remove the core or seeds and fibers. Then slice the pepper into strips 1/4 inch wide and as long as the pepper. Spread the strips on the dryer rack and leave for 8 to 10 hours. The strips will darken and become ropy or leathery when they are ready to store.

Potatoes. Irish or white potatoes dry well, too, if you wash and peel, then slice the potatoes into 1/8 inch or slightly thicker sections. Blanch in

Fig. 7-2. You can dry corn on the cob in a dehydrator simply by placing the corn ears on the rack and putting the rack in the dehydrator. You may need to turn the corn ears once or twice during the drying period. When dried the kernels will tend to shrink and pull away from each other. At this point the corn can be sealed and stored.

boiling water for 5 to 7 minutes, and then drain. Next, spread the potatoes on the racks and dry until the slices are flexible but very leathery. You should turn the slices every 2 or 3 hours and allow about 15 hours for the slices to dry completely.

Pumpkins. For an astonishing experience, choose a pumpkin and peel and chop into cubes 1 or 2 inches square, then blanch for about 2 minutes and drain. Spread the cubes on the dryer racks and watch them shrivel away to a fraction of their former size. You can store a huge pumpkin, after it is dried, in a coffee can easily. It takes the cubes about 12 hours to dry to a firm but pliable toughness.

Radishes. Radishes dry easily and quickly if you pick young ones, clip stem and roots off, wash, and slice into thin slices. Dry for about 8 hours. You may want to turn the slices occasionally and perhaps rotate the racks inside the dryer.

Summer Squash. Summer squash makes a nice, crisp "cracker" to eat with various party spreads or dips. Pick young squash or zucchini which have not developed large seeds. Wash and then cut off the blossom and stem ends. Slice the squash into the usual 1/4 inch thicknesses, and spread to drain. Then place on racks and allow to dry until crisp. Turn every four hours and leave the slices in the dehydrator for about eight hours, slightly more if necessary.

Winter Squash. Winter squash can be dried much like pumpkin. Peel, chop into cubes, and immerse into boiling water for about 5 to 10 minutes. Spread on racks and allow 12 hours or more to dry.

Sweet Potatoes. Sweet potatoes can be dried much like Irish potatoes. First, wash and peel the potatoes and slice into 1/4 inch slices. Cook in boiling water for 5 minutes or slightly longer, drain, and then spread on the racks to dry. Wait until the slices are light and brittle before you remove them. It may require 15 hours for the slices to become dry enough to store.

Turnips. To dry turnips, follow the instructions for drying radishes, but after turnips are sliced, blanch in boiling water for 3 minutes, drain, and then spread on the drying racks. It usually takes turnip slices 10 to 12 hours to dry completely.

Other Drying Methods

Obviously, if you are solar drying, you will need two or three times longer to dry most of the neutral vegetables listed here. Oven drying is usually much shorter in time. The predrying preparations are the same.

ROOT CELLAR STORAGE

Many of the vegetables listed in the canning and dehydrating section of this chapter can be stored for fairly long periods of time in a root cellar or other storage place that is cool, dry, and protected from insects and rodents.

How To Store Vegetables

Beans, for instance, will keep for weeks once they are dried—either on the vine or in a sunny spot on the porch—and kept away from moisture. Once they get wet they will sprout and then they should be discarded, unless you want to cultivate sprouts.

The beans can be kept in a sack that allows air to circulate if the sack is suspended from a rafter or nail. Moisture will not condense at the bottom of the sack like it would if in contact with the floor.

You can keep beets for shorter periods of time, depending upon the moisture level and temperatures of the storage areas, by leaving them in the ground until the last possible day, then pulling them with stalks and leaves still attached. Tie the beets in bunches and hang them from a nail or along a rope or heavy cord. Do not allow them to get wet.

Cabbage can be placed in a sack that has a loose type of fabric so air can circulate, and the sack can be hung in a dry storage area. If, however, the cabbage heads get wet and decay, you will notice the fact long before you visit the root cellar or basement. The odor is quite powerful and decidedly unpleasant.

Carrots can be hung up, preferably with the stems still on them. They will keep for several weeks, though they will not have quite the taste of fresh carrots. You can also bury the carrots, turnips, and other root crops in loose sand, as long as there is not a danger of dampness.

There are many ways of drying corn, but one of the oldest and most unusual methods of storage is not done in a root cellar at all; the corn is left in the field and on the stalks. However, you cut the stalks and stack them in a shock. Leave the ears still on the stalks, and, as you need the ears, use the ones on the stalks nearest the outside ranges of the shock. As you work your way into the shock, you will find that the corn in the deepest parts of the shock is still ripe and ready to use, but it is in a dehydrated state. The problems here are excessive moisture, rodents, and birds. You could store the corn in a basket or crib, and it will be good for weeks, even months.

Many people pull pepper plants and hang them upside down. The peppers ripen and remain edible for weeks with no preservative measures. The same idea is used with considerable success with other vegetables, notably tomatoes. You can also string peppers together with needle and thread, with the thread passing through the stem part of the pepper. Hang the strung peppers on a nail or on a wire or cord.

Pumpkins can be stored in a root cellar or simply on the floor of a vacant room, as long as the temperatures and moisture level remain at a normal point. You don't want the pumpkins to freeze, for instance, but they will last for weeks, sometimes months, if they are left intact. If you cut one, however, or if one is damaged in handling, the decay will begin instantly in that particular spot.

Vegetable spaghetti, which can be dried in a dehydrator or solar dryer, can be stored without treatment in a root cellar or on the porch floor—wherever you can store pumpkins.

Winter squash will keep for long periods of time under similar conditions.

The root cellar is an excellent place to store many types of low-acid or neutral foods. The beauty of this type of storage is that you don't need to pay one cent for any type of energy to process or to preserve the food. Of course, you will lose some of the food, but you lose some by freezing, canning, and dehydrating, too.

How To Store Potatoes

Of all the foods that keep well in the root cellar, however, few if any surpass potatoes.

You can spread bushels of potatoes on the earthen floor or on shelves or wood floor—concrete is bad for almost any kind of food not in containers—and the spuds will keep until the following spring. You may want to sprinkle a little lime over the potatoes, but this treatment is not essential.

We always dig our potatoes as soon as the plants start to die. We have found that if we leave the potatoes in the rows, they begin to decay as soon as the next heavy rain saturates the soil. After we dig them, we spread them on a sheet of plastic or even on the grassy areas near the house and hose them down to remove excess dirt. Then we spread them in an open area, but not where the sun can hit them directly, where they can air dry for several days. Incidentally, the sun will cause green spots on the skins of potatoes, and these deep green areas are generally concentrations of solanum, a very dangerous poison. As perhaps nearly everyone knows, the potato skins, as well as the flesh, contain solanum in minute quantities. Cooking destroys the toxic qualities of the solanum-laced potatoes, so there is no danger for human consumption. Even if you eat the meat of the potato raw there is very little, if any, danger. In these greenish concentrations, however, there can be significant amounts of poison present.

Don't be alarmed. Leaves of cherry, peach, and apple trees also contain toxic substances, and the plants of potatoes and tomatoes themselves contain poisonous substances.

After the potatoes have dried for several days, take them to the storage area and spread them out to cure. As you store them, the potatoes will have skins so thin that you can rub it off with pressure from your thumb, but as they dry, the potatoes develop thickness and leathery qualities that protect the tubers from decay and excessive insect damage.

Fig. 7-3. You can also freeze any of the vegetables discussed in this chapter. You will need to blanch the vegetables that have already been washed and prepared for processing. When the blanching is completed, drain the vegetables and then pack in freezer bags or cartons. To freeze corn on the cob, immerse several ears at a time in boiling water. Use tongs to remove the corn, and then lay it on a platter to cool. Drain, then pack in freezer containers.

We also gather all of the tiny potatoes that we see when we are digging. Even potatoes not as large as a golf ball are taken to the storage area. Then, when the spring planting season arrives, we use these tiny spuds as seed potatoes. Since they are small we do not need to cut them, and since they are not cut there is no need to allow a day or so for the cut to cure or heal. The marble-sized seed potatoes germinate superbly and produce potatoes as nice as any we have ever grown from the potato eyes of seed potatoes.

You can also store sweet potatoes for months in the dry storage area. We usually do nothing except check once in a while to see that there are no problems.

There are ways of storing food other than by the methods already suggested (Fig. 7-3). Chapter 8 deals with making pickles and relishes, some of the tastiest foods that can come from a garden.

8

Making Pickles and Relishes

Pickles and relishes are among the most delightful of foods. Pickles also are among the few foods that can be preserved well through a process of fermentation.

Another interesting and delightful aspect of making pickles is that, at least on occasions, you use parts of the food that would otherwise be thrown away. Watermelon and other melon rinds, for instance, can be pickled, and the results are delicious.

Since there are almost as many types of pickles as there are social groups and nationalities, in this chapter we shall discuss the four primary groups of pickles and relishes. Generally, whatever principles hold true for one food in the group will be true for others, unless noted otherwise, in the family.

Dozens of kinds of foods can be pickled. Among the favorite vegetables are tomatoes, cucumbers, peppers, beans, corn, cauliflower, onions, okra, and beets. Some melon pickles can be made easily. Several fruits, among them apples, peaches, pears, crabapples, plums, some grapes, and figs can be preserved by pickling. Eggs and certain meat products also work well as pickled foods, as do a few types of fish.

The four major types of pickled foods are brine pickles, fresh-pack pickles, fruit pickles, and relishes.

BRINE PICKLES

As the name implies, brine pickles are made usually from a salt base or salt soak. When you buy salt, do not buy table or iodized salt; instead buy pickling salt, which is readily available from any canning supply store.

There are several types of pickles that can be made through short methods; other pickles require several days of work before the pickles are ready to store. You can also make sweet pickles, dill pickles, and sour pickles through the brine method.

Ingredients

The vinegar you will need should have an acidity

level of four to six percent, and it is always better in such efforts to purchase a good name brand of vinegar. You can use either cider vinegar or white vinegar for your pickles, but use white whenever you want pickles that are very light in color.

If you use too much salt, excessive sugar, vinegar of the wrong acidity level, or the wrong kind of salt you may get pickles that are tough, wrinkled, or distasteful. If you cook pickles too long they may become mushy instead of crisp.

Your basic steps in making pickles are food selection, preparation, soaking or brining, clearing or desalting, processing, and storing.

Always try to use vegetables that are young, firm, and fresh. Vegetables that are too mature, soft, old, or unsound will not yield high-grade pickles. If you are pickling tomatoes, you will want to use green tomatoes rather than ripe ones. Squash should be young and tender; okra must be picked young and tender.

Sweet Pickles

To make sweet pickles, you may use any of several methods, but in essence the major difference is, as the name suggests, the amount of sugar used. You will use sugar in nearly all kinds of pickles, but sweet pickles have a greater concentration.

FRESH PACK PICKLES

Several methods of making sweet pickles from a variety of vegetables follow.

Cucumber Pickles

After you pick and wash your cucumbers, put them in a large vessel—preferably a plastic container or a heavy crock rather than a metallic pot—and cover them with cold water. Put a lid on the vessel and be sure that all the cucumbers are under the water. Let them soak all night.

The next day make a syrup of two cups of vinegar, 1 cup of water, 1/2 teaspoon of mustard seed, 1 cup of granulated sugar, or slightly less, and 1 teaspoon of various pickling spices. You can buy a box of mixed spices, and these will work well.

Bring the syrup to a boil and slowly add the cucumbers to the mixture. After the liquid comes back to the boiling point cook for 3 to 5 minutes. You can tell when the cucumbers are ready because they will start to look pale.

Pack the cucumbers into sterilized jars and add a scant teaspoon of salt to each jar. You may use onion slices if you like or peppers, or you may omit all of these. Cover the cucumbers with boiling syrup and seal.

You may serve the pickles the following day, if you like, but the longer the pickles are left in the syrup the stronger flavor they will have. The pickles will keep for over a year after they are sealed. If you would prefer sweeter pickles, in your next run increase the sugar content by 1 to 2 cups.

A variation of the sweet cucumber pickle uses 9 pounds of cucumbers sliced lengthwise. Soak these slices in ice water—so cold that it is actually slightly slushy—for 3 1/2 hours. Use 1 gallon of water, slightly more than 1 cup of salt, and 5 cups of sugar. You may vary the amount of sugar and salt to suit your own taste.

Bring the mixed liquid to a boil and then, while the syrup is still piping hot, pack the drained cucumbers into sterilized jars. Pour the boiling liquid very slowly over the cucumbers until the slices are covered. Seal the jars and store for 2 weeks before you sample the pickles. You may add tiny pearl onions or onion slices if you prefer.

Bread and Butter Pickles

A delicious bread and butter pickle can be made starting with 1 gallon of small cucumbers. Wash but do not peel, then slice crosswise in very thin slices—paper-thin if you can manage it—and add 8 small, thinly sliced white onions and 2 green peppers, also sliced into thin strips. Mix 1/2 cup of salt with the vegetables and cover with crushed ice. Let the mixture stand for 3 hours and then drain thoroughly.

Make a mixture of 5 cups of sugar, 5 cups of vinegar, 1 1/2 teaspoons of turmeric, 1/2 teaspoons of ground cloves, 1/2 teaspoon of celery seed, and 2 tablespoons of mustard seed. Pour the mixture over the vegetables and place on low heat to bring the contents of the vessel to the scalding point, but do not boil. Fill sterilized jars with the mixture and seal.

If you prefer, you may omit either the onions or the peppers, or both, and you will still have a delicious pickle.

Frozen Pickles

If you want to try frozen pickles, which are extremely crisp and tasty, use 6 cups of sliced cucumbers and a mixture of 2 cups of sugar, 1 tablespoon of salt, and 1 cup of white vinegar. Mix sugar, salt, and vinegar. Pour the mixture over the cucumbers and let stand overnight. If you like, you can add small white onions and green peppers—1 cup of each, thinly sliced—to the mixture, and you can add a teaspoon of dried dill weed if you want a stronger flavor.

The next day fill a plastic container with the vegetables and pour the vinegar mixture over them. You can keep these pickles in your refrigerator for several weeks, or you can freeze them and keep the pickles indefinitely.

Speedy Pickles

If you are in a hurry to put up cucumbers, here is a speedy pickle that takes only a few minutes. Start with 2 cups of sliced cucumbers, 1 small onion sliced thin and separated into rings, 1/2 green pepper (optional), 1/2 cup of vinegar, 1/2 cup of sugar, 1/2 teaspoon of mustard seed, and 1/2 teaspoon of celery seed.

Put everything in a container and cook on high in a microwave oven for 7 minutes, or you can put the ingredients in a pot and cook on high on a traditional burner for 15 minutes. Then chill the mixture and pack into sterilized jars.

If you want more pickles, simply increase all of the ingredients proportionately. For instance, use 4 cups of sliced cucumbers and then double all of the other amounts. The cooking time, however, remains the same.

Other Variations

For interesting variations on the above recipes, you may add a number of ingredients to vary the flavors. To 8 cups of sliced cucumbers add 1/4 cup of raisins and then proceed with an ordinary syrup mixture. You can also add a teaspoon of curry powder, sections of cauliflower, or virtually whatever vegetable suited for pickling that you prefer.

If you want sour pickles, cut the sugar totally or reduce the amount significantly. Instead of 3 cups, cut the amount to 1 cup, for instance.

Dill pickles can be made simply by adding dried dill weed (for appearance as well as for taste) as desired. A small bunch of dill can be placed in each jar before the pickles are added, or you can insert it afterward. You can vary the amount to suit your own needs and tastes.

Squash Pickles

People who do not try squash pickles are missing out on one of the best pickles you can make. For large amounts—and since squash is usually so plentiful in the summer, why not make several jars—start with 4 quarts of squash sliced very thinly. Add 2 quarts of onions, also sliced thinly. Next you need two quarts of crushed ice.

Mix these 3 ingredients, pour over 1/2 cup of salt, and cover the entire mixture for 3 hours. After the soaking is completed, drain and pack the squash and onions into quart jars.

Mix 5 cups of sugar, 5 cups of vinegar, 1 1/2 teaspoons of turmeric, 1 teaspoon of mustard seed, and 1 teaspoon of celery seed. After the mixture is ready, pour into a pot and place on a burner. Heat to the boiling point but do not allow it to boil. Stir constantly during the heating period.

Pour the mixture over the squash and onions which you have already packed into quart jars. Seal the jars and wait for 2 weeks before sampling. These pickles keep very well for a year or even much longer, and they remain very crisp and firm.

The best squash to use are young yellow or crookneck squash, but you can experiment with zucchini or other types of squash. Do not, however use winter squash.

Green Tomato Pickles

When the end of the summer comes and you have

tomato plants loaded with green fruit that cannot possibly ripen before a killing frost, you can wrap the larger tomatoes as directed in an earlier chapter and allow the fruit to ripen inside the newspaper wrapping. You need not waste the smaller tomatoes either.

You can pick 1 gallon of small green tomatoes, slice these 1/4 inch thick, and sprinkle 1/4 cup of salt over the slices. Allow the salt and tomato mixture to stand for 8 to 10 hours. Drain and pour 2 quarts of boiling water over the tomatoes and let this mixture stand for half an hour. Then drain, cover again with cold water, and drain again.

Mix 3 cups of vinegar, 1 cup of water, 4 cups of sugar, 1 tablespoon of mixed spices, 1 tablespoon of celery seed, 1/2 teaspoon of allspice, and 1 tablespoon of mustard seed, and bring this mixture to a boil. Pour the syrup over the tomatoes and let stand for another 8 to 10 hours. Then bring to a boil again. Pack the pickles into jars, cover with the syrup you mixed, and seal the jars.

You can also add tiny onions to the pickles if you like. Green peppers mix well with this type of pickle as well. If you prefer you can tie your spices into a small bag and then remove them before you seal the pickles, since spices tend to cause pickles to darken more than many people like.

You can also add other spices, such as cinnamon and cloves, if you wish. Use 1 tablespoon of cloves or cinnamon.

Pickled Beans

Start with 4 pounds of green beans that are fresh and firm, yet tender. Wash and string the beans and then cut or break into 1 or 2 inch lengths. For each pint measure, you may add the following ingredients for extra flavor: 1 clove of garlic, 1 dill head, and 1/4 teaspoon of hot pepper. Combine 6 cups of water, 6 cups of vinegar, and 1/2 cup of salt, and bring to a boil. While waiting, pack beans into jars, leaving 1/2 inch of space. Pour the boiling liquid over the beans and boil the jars for 10 minutes in a water-bath canner. You should wait at least 2 weeks before using beans to allow the flavor to fully develop.

If you desire a sweeter, spicier bean, place precooked beans in jars and cover with a boiling mixture made of 6 cups of cider vinegar, 1 cup of brown sugar, 1 teaspoon of salt, 1/4 teaspoon of pepper, 1/2 teaspoon of allspice, 1 stick of cinnamon, and 1 tablespoon of whole cloves. Adjust spice measurements for your individual tastes.

Pickled Okra

Small, tender pods of okra make the best pickled okra. To prepare okra, wash and drain each pod, trimming the stem end. Prick each pod and pack into jars. Into each jar you may add 1 small bunch of dill and 1 garlic bud for extra flavor. Hot pepper adds significantly to the taste. Combine and bring to a boil 2 pints of white vinegar, 2 quarts of water, and 2/3 cup of salt. Cover packed okra with the boiling brine and boil the jars in a water-bath canner for 10 minutes. Wait several weeks before using to allow flavor to fully ripen.

Watermelon Rind Pickles

When you have finished eating the watermelon, do not throw the rind away. You can trim off all of the pink and green parts (the peel and the base of the meat) and you have a pale green section left. Cut up this rind into 1-inch cubes and soak it 10 hours in a solution of 8 tablespoons of salt to 1 gallon of water, or you can reduce the amount to 4 tablespoons of salt to 1/2 gallon of water if you don't have much watermelon rind.

After the soaking, drain the rind and cook it until it is not quite tender. Drain, and then make a syrup of 4 cups of sugar, 2 cups of vinegar, 4 teaspoons of whole cloves, 8 cinnamon sticks, and a sprinkle of mustard seed. As usual, you can use a piece of cloth and tie all of the spices in a small sack so that they don't stay in and darken the pickles.

Heat the syrup to the boiling point, but do not boil. After the syrup cools for 10 to 15 minutes add the watermelon rind and cook until it becomes almost transparent. Then pack into hot and sterilized jars and seal.

You can add 1 slice of lemon, ginger root, or any other spices that you enjoy.

Pickled Beets

Start with young beets that have not been exposed to sunlight for a prolonged period. Leave the roots and 1 inch of the stem on and boil the beets until the skins slip off easily. Remove skins, stems, and roots, and slice or quarter the beets. If they are small, you can leave them whole.

Make a syrup of 2 cups of sugar, 2 cups of water, 2 cups of white vinegar, 1 teaspoon of cloves, 1 teaspoon of allspice, and 1 tablespoon of cinnamon. Pour this mixture over the beets and boil all of the ingredients for 10 minutes. Pack the beets into sterilized jars and cover with liquid and seal immediately.

If you have a large quantity of beets to preserve at one time, you may wish to can beets plain and later open a jar for use as pickles. To make pickled beets of your canned beets, open a pint jar—or whatever amount you need—and drain the liquid from the beets. Cover the beets with a syrup made of 1/2 cup of sugar, 1/2 cup of water, 1/2 cup of vinegar, 1/4 teaspoon of cloves, 1/4 teaspoon of allspice, and 1/2 teaspoon of cinnamon. Boil for 10 minutes. Return the beets and syrup to a jar and let the beets stand in syrup for a day or two before using. Refrigerate beets when cooled.

Pickled Vegetable Mixture

Select your favorite vegetables, such as cauliflower, carrots, onions, peppers, and green tomatoes. Scrape the carrots with the edge of a sharp knife until they are "peeled" and chop the cauliflower into small bite-sized chunks. You may halve tomatoes or leave whole. You can use as many vegetables as you like, or you can use only one if you wish.

Make a syrup of 1 quart of white vinegar, a small bag of whole cloves, allspice, cinnamon, and other spices that you like (1 tablespoon of each), and 4 cups of sugar. Heat the syrup to boiling and then pour over the vegetables and let the mixture and vegetables stand for at least 10 hours.

Then bring back to a boil and hold it there for another 5 to 10 minutes. Remove the bag of spices, pack the vegetables into hot, sterilized jars, and close. Process 10 to 20 minutes in a water-bath canner and then store for 3 weeks before eating.

Pickled Peppers

Select mature peppers. Wash, drain, and make a 1-inch slit in each side of the peppers. Make a brine mixture of 1 1/2 cups of salt, 4 garlic cloves, or 2 if you want a milder taste, 10 cups of vinegar, 2 cups of water, and 1/2 cup of sugar. Make sure the salt is well dissolved, then pour the mixture over the peppers—which should be a gallon or so of mixed peppers. Let the brine and peppers stand overnight, and then rinse and drain.

At this point put in the other ingredients and cook for 15 to 25 minutes, but do not cook until peppers are mushy; you want them to be firm and crisp. Take out the garlic, pack the peppers into hot, sterilized jars, and cover with the brine mixture.

Close the jars and process for 10 to 20 minutes in a water-bath canner.

Caution: If you are preparing hot peppers, be sure to wear rubber gloves. The contact of the hot peppers with your flesh may not be uncomfortable at the time, but in about an hour or so a severe irritation may appear and cause great discomfort for several hours. Do not touch your eyes until you have washed your hands thoroughly or until you have removed the rubber gloves. Exert great caution when you are making the slits in the peppers since a tiny drop of juice may contact sensitive areas of your skin or eyes and cause pain for several minutes.

Pumpkin Pickles

To make pickled pumpkin, start with about 12 pounds of good, firm pumpkin. Do not use pumpkin that is soft or overripe; the meat should be crisp and hard. Weigh the pumpkin before you start the rest of the processing, since the ingredients are based on the proportion of uncooked pumpkin to the condiments.

Peel the pumpkin and scrape away all of the pulpy matter from the center. You may need to cut down below the seed cavity pulp for 1/2 or 2 inches in order to get to the really firm meat.

When you have the pumpkin ready, cut it into long strips by following the natural grooves with a sharp knife, but be very careful. Pumpkin flesh is very hard but subject to sudden splitting, and you may find that the knife blade will glide suddenly through the meat of the pumpkin, slicing your hand before you can control the blade.

Once you have sliced the pumpkin into large strips, chop it into 1/2 inch chunks. Cover the chunks, which should be about 1/2 inch thick, 1 1/2 inches long, and 1/2 inch wide, with a generous coating of sugar. Allow the pumpkin to stand overnight.

Add 6 cups of vinegar, 3 or 4 tablespoons of whole cloves, and a dozen sticks of cinnamon. Put the mixture in a cooking vessel and cook over medium heat until the pumpkin starts to become transparent. The syrup will also start to thicken.

Use a large spoon, preferably a wooden one, to scoop the pumpkin into hot, sterilized jars and cover with the syrup. Close the jars and process the pickles in a water-bath canner for about 15 minutes. Store the pickles for at least 15 days before opening.

Pickled Eggs

Pickled eggs are a good old-fashioned food in many parts of the country, but they are especially popular in the Appalachian territory.

Place a dozen eggs in a pot of water and allow them to sink to the bottom of the pot. If any egg floats or tends to float, replace it, since this egg will not be suitable for use in food of any sort. Turn the burner on high, and, when the water has reached a rolling boil, maintain it for at least 3 minutes. Then turn the heat down to simmer and allow the eggs to cook for 15 more minutes. You want the yolks to be very firm for pickles.

If your eggs are very fresh, you will need to take them from the hot water and run cool water over them for 2 to 3 minutes and then place them in a vessel of very cold water. You can also allow the eggs to air cool for 2 minutes, then put them straight into the cold water for 10 minutes.

If the eggs are several days old, you may omit this step completely.

While the eggs are cooling, make a syrup of 2 cups of sugar, 2 sticks of cinnamon, 1 tablespoon of allspice, 1 1/2 teaspoons of salt, 3 1/2 cups of vinegar, and 1 1/2 cups of water. Heat the syrup mixture until all of the ingredients are dissolved except for the spices, and pour the syrup over the peeled hard-boiled eggs.

Close the jars and do not seal. You can keep the pickled eggs in the refrigerator for several days. You should let the eggs pickle for 2 days before sampling.

FRUIT PICKLES

Many kinds of pickles from fruits can also be made. Here are some suggestions.

Pickled Peaches

Pour 2 to 3 quarts of water into a saucepan and add 10 cups of sugar, 7 cups of vinegar, 12 two-inch sticks of cinnamon and 1 1/2 tablespoons of cinnamon. Bring to a boil and cover. Let simmer while preparing the peaches.

Wash the peaches and dip them in boiling water for 1 minute, then in cold water to make the peach peel easier. You will need a generous peck of peaches for the above amounts of ingredients. Halve the peaches and remove the seeds. Place the halves into cold water mixed with 2 tablespoons of salt and 2 tablespoons of vinegar per gallon to prevent the halves from turning dark. Drain just before you are ready to use them.

Add the peach halves to the boiling syrup—you will not be able to do them all at one time—and continue to heat for 5 minutes. Pack hot peaches into jars and cover with the boiling syrup. Add a stick of cinnamon and some cloves if desired. Process jars in a boiling water bath for 20 minutes.

Spiced Fruit Pickles

For these pickles you may use apples, peaches, pears, or similar fruits. Make a syrup of 5 cups of brown sugar, 2 1/2 cups of vinegar, 2 or 3 cinnamon sticks, and several whole cloves—several tablespoons if a strong clove taste is desired. Cook the mixture until all the sugar is dissolved and the clove taste has a chance to permeate the mixture.

About 15 minutes will be sufficient.

If you are using peaches, halve them and remove the pits. If you use apples or pears, use a corer or a thin knife blade and cut out the core area. Then slice rings from the fruit—peeling is preferred but optional—and drop the rings into the syrup. Cook until the fruit can be pierced easily with a fork and slight pressure.

After the tender point of the fruit has been reached, pack the fruit into sterilized jars that are already hot. Heat jars by pouring very hot water into jars and let water remain in them until ready to use them. Cover the fruit with the hot syrup.

Seal the jars immediately and store the pickles for 2 weeks before using.

Fig Pickles

Select young figs and wash and drain, then place them in a pot and cover the figs with boiling water. Allow them to stand in the water for 10 minutes. While the figs are soaking, make a syrup of 2 cups of water, 12 cups of sugar, 2 1/3 cups of vinegar, if you prefer a truly sour taste; only 2 cups if you prefer a milder taste, and a small cloth bag filled with assorted spices. You may use cloves, cinnamon, and any whole spices that you like. Use 3 tablespoons of spices but remove the bag and the spices before sealing.

Cook the figs—to the boiling point and hold for 10 minutes—every 15 to 24 hours for 3 days. At the end of the third day, pack the figs and syrup into sterilized jars and process in a water bath for 15 minutes. Allow the pickled figs to stand for 2 weeks before opening.

Variations

The recipes given in this chapter are fairly basic approaches to making pickles. You can feel free to add or subtract ingredients from most of the brines or syrups according to you own personal preferences, but do not cut back significantly on the salt and vinegar, since these two ingredients are the preservatives in the foods.

You need not be limited to the foods mentioned here. You can experiment with a variety of fruits and vegetables and make your own recipes for pickles as well as relishes.

RELISHES

Generally speaking, a ***relish*** is any of a variety of foods served with a meal to give flavor or as an appetizer. Relishes can be made from a wide variety of vegetables, such as corn, cabbage, tomatoes, various fruits, and combinations of fruits and spices.

Apple Relish

To make an apple relish, you will need sweet green peppers, sweet red peppers, hot peppers, onions, apples, celery, and brown sugar. First, wash the vegetables and cut out any bad spots. Then, if you have a food chopper with a coarse blade, run all the food through the chopper.

You need 6 medium sweet green peppers and the same amount of sweet red peppers. The hot peppers, two of them, can be small to medium size, unless you want a very tangy and hot relish. Six medium or 5 large onions will suffice. The 6 apples you need should be large, but if they are medium, you can use 7 or even 8. One large stalk of celery is needed, and the 1 cup of brown sugar should be packed rather than loosely measured. One or 2 clove spices can be added whole.

If you don't have a food chopper, use a kitchen knife to cut the vegetables into small pieces. Cook the vegetables first until they are tender, and then cover them with white vinegar and stir in the brown sugar. Continue cooking for another 5 to 10 minutes—until the vegetables are tender enough to serve—and the add the chopped apples. Cook for 10 more minutes or until the brown sugar turns the entire mixture an amber color. Pack into jars and process for 10 to 15 minutes in a water bath. Do not use a pressure canner, since the pressure will cause the relish to become mushy and overdone. The relish will be ready to eat in a week.

Apple Chutney

Peel, core, and chop a dozen sour apples—the early summer variety or late apples not fully ripened. Add

2 chopped red peppers, 2 cups of granulated sugar, 6 tablespoons of lemon juice, 3 chopped sweet green peppers, one cupful of chopped raisins, 1 tablespoon of cinnamon, a dash of salt, and slightly more than 1 pint of white or cider vinegar.

Cook the entire mixture together until it is thick and dark-colored. All the vegetables and fruit should be tender. Then pack into hot, sterilized jars. Close the jars and process for 15 minutes in a water-bath canner. Eat when you are ready.

Beet Relish

Use 1 quart of beets, finely chopped, 1 small head of cabbage chopped, 1 cup of chopped sweet red peppers, 1 tablespoon of salt, 1 1/2 cups of granulated sugar, and 3 cups of vinegar.

Cook all the ingredients together until they reach the boiling point, then lower the heat and cook for about 10 more minutes, until everything is tender. Pack into hot, sterilized jars, close, and process in a water bath for ten minutes. The relish can be eaten immediately.

Chow-Chow

Chop about 10 to 12 pounds of green tomatoes, 8 large onions, 12 green bell peppers, and 6 hot peppers. Add enough salt to cover the mixture, and allow it to stand for 8 hours.

Drain, add a small bag of spices—1 tablespoon each of cinnamon, cloves, allspice, and 3 tablespoons dry mustard. Next, add a quart of white vinegar and 2 cups of sugar. Cook until tender after bringing to a boil and reducing heat. Simmer for at least 15 minutes. Finally, pack the food into hot, sterilized jars, close, and process in a water bath for 15 minutes.

Chow-Chow with Cabbage

Chop by hand or run through a food chopper 1 pint of sweet red peppers, one pint of sweet green peppers, 1 pint of green tomatoes, 3 green peppers, 1 pint of onions, and 2 pints of cabbage. Cover with salt and let stand overnight. You should measure the vegetables after chopping.

The next day drain, then add 2 tablespoons of celery seeds, 1 teaspoon of mustard seeds, 1 two sticks of cinnamon. Stir in 4 cups of sugar and 1 quart of vinegar. Bring the mixture to a boil and cook until all the vegetables are tender and blended well. Pack into hot, sterilized jars and close, then process for 15 minutes. If you wish, you can leave 1 jar open and eat it raw.

Corn Relish

You will need 8 cups of fresh corn kernels. If you start with corn on the cob, cut the kernels off the cob and leave them uncooked temporarily, or you can cook the corn on the cob for 5 minutes before cutting off the kernels.

Next, chop up 2 cups of red sweet peppers, 2 cups of green sweet peppers, 1 1/2 cups or more of chopped celery, and 1 cup of chopped onion. Add 1 tablespoon of salt, 1 teaspoon of celery seed, 1/2 teaspoon of turmeric, and 2 teaspoons of dry mustard.

Cook the corn for 5 minutes, then add all the other ingredients. Mix in 1 quart of white vinegar, and 1 1/2 cups of sugar. Cook together until all ingredients have mixed well and all vegetables are tender—about 15 minutes of boiling heat. Pack into hot, sterilized jars and process in a water-bath canner for 15 minutes. The relish is ready to eat or store at this time.

Cherry Relish

Pit and wash 2 cups of sweet or sour cherries and chop 1/2 cup of raisins. Get 1 full cup of chopped walnuts, pecans, or other nuts—a mixture of these is fine—and a scant 1/2 teaspoon of such spices as cloves.

Cook the cherries and raisins in 1/2 cup of brown sugar, 1/2 cup of honey, and 1/2 cup of vinegar, and simmer for 45 minutes. You can add the nuts just before you take the mixture off the heat. You can also add one sweet green pepper. Cook for 5 minutes more and remove from heat. Pack and close. Process for 10 minutes in a water-bath canner. The relish is ready to eat after one week.

Pear Relish

Start with 12 pounds of pears. Peel, core, and chop. Chop 4 onions and 6 green sweet peppers. Add, if you like, 1 cup of nuts and 2 stalks of celery. If you plan to process the food omit the nuts.

Mix the food into a syrup made of 4 1/2 cups of vinegar, 1/2 tablespoon of salt, 3 1/2 cups of sugar or 2 1/2 cups of strained honey, and 1 tablespoon of allspice.

Let the mixture stand for 10 hours, then pack it into jars and close. Process for 20 minutes in a water bath.

Piccalilli

Chop 1 quart of cabbage and 1 quart of green, firm tomatoes (measure after chopping). Add 4 red sweet peppers and 4 sweet green peppers. Sprinkle about 1/4 cup of salt over the mixture and allow it to stand for ten hours. Then drain in a sieve for at least an hour. If you prefer, you can add two or three large onions to the mixture before allowing it to set overnight.

Make a syrup of 2 cups of white vinegar, 1 cup of water, and 2 cups of brown sugar. Then add a teaspoon each of cloves, dry mustard, turmeric, and celery seeds. Boil the mixture until the sugar and spices are fully dissolved or at least mixed well with the liquids.

Add the chopped vegetables and bring to a boil, then cook for 3 more minutes. Remove from heat and pack into hot, sterilized jars, and process, after closing, in a water bath for 10 minutes.

You can, if you prefer, add a couple of hot peppers, in addition to, or in place of, the other peppers.

Pepper Relish

Wash and drain 12 sweet green peppers and 12 sweet red peppers, plus, if you prefer, 3 hot red peppers. Chop these peppers finely and then add 2 large chopped onions. Cover with boiling water and let the mixture stand for 1/2 hour, then drain. Add 1 quart of vinegar, 2 tablespoons of salt, and 2 cups of sugar.

Bring the entire mixture to a boil and then reduce the heat and cook for 20 minutes. Pack the food into hot, sterilized jars and cover with the hot liquid. Close the jars and process for 10 minutes in a water-bath canner.

Tomato Ketchup

For a good homemade tomato ketchup, start with 1 cup of vinegar and add 1 1/2 teaspoons each of whole cloves and cinnamon, 1/2 teaspoon of allspice, and 1 teaspoon of celery seeds. Bring to a boil and remove from heat.

Next chop up about 30 medium tomatoes—you can use tomatoes that are really not good enough for canning—and add 1 cup of chopped onions and 1 teaspoon of red pepper.

Cook the vegetables and stir constantly for 20 minutes. After the cooking is completed, put the mixture through a sieve and cook with 1/2 cup of sugar until the volume is reduced by half.

Now strain the vinegar so that the spices are removed and add 4 teaspoons of salt. Stir the vinegar into the cooked tomatoes and then boil until thick. The final cooking requires about 1/2 hour. Pour the ketchup into hot, sterilized jars and seal.

Another Version

For another version, start with a gallon of tomatoes that have been chopped and cooked until pulpy. Be sure that the tomatoes have been cooked until they are fully tender. Run the tomatoes through a colander, add 2 tablespoons of salt, 2 tablespoons of black pepper, 4 tablespoons of sugar, 1 tablespoon of mustard, 1 pint of vinegar, and a spice bag containing cloves, cinnamon, and any other spices that you like. You may also add one scant cup of minced sweet red peppers.

After all of the ingredients are in, cook the ketchup for 1 1/2 hours or until the mixture thickens. Then pour the food into clean, hot jars and process for 15 minutes in a water-bath canner.

You can also add onions, hot peppers, nutmeg, ginger, and other preferred spices, and you can add a spoonful or two of flour if you need more thickening. By the same token, you can omit several ingredients if you need to watch your food intake.

For instance, you can omit the sugar or at least cut the amount back drastically. You can also leave out the salt if you plan to use the ketchup quickly rather than store it for long periods of time.

Variations

You will find that if you experiment with various types of relishes that you can come up with your own personalized versions that are more pleasing to you than any ready-made recipe can be. For instance, you might want to seal half the recipe as it is described here, but in the remaining half tailor the taste to you by adding more hot peppers, less hot peppers, more or less spice, different types of spices, more or less sugar, any other change or substitution that can be made without damaging the recipe severely.

You are likely to find, too, that many of the relishes that you do not enjoy from the commercially prepared jars in supermarkets are delightfully delicious when made from your home-grown vegetables. Many people, until they start preserving their own food, do not really know how flavorful canned, frozen, or dried food can be, and the same holds true for relishes and chutneys, because you pick the fruit at the peak of its ripeness, and cook and season it as you like it.

We have found, for instance, that store-bought canned beans and tomatoes, as well as many other prepared and preserved foods, are unpalatable to us. We cannot eat canned biscuits from the store shelves, and we have a very difficult time digesting any prepared dessert that is sold in stores. Even our child and our family pets will steadfastly refuse to eat many of the bought foods, since they have been accustomed to home-preserved preparations.

9

Soups and Sauces

If there is such a thing as a universal food, it might well be soups. Soups can be made easily and quickly; they can consist of only one major ingredient, whether vegetable or meat, and they can contain all vegetarian ingredients, or virtually all meat, or combinations of meat and vegetables.

Soups may be composed of the solid foods plus water, milk, or meat-stock base. You can eat them hot or cold, and they can be made for immediate use or canned, frozen, or dried for future use.

One of the most attractive aspects of soups is that they can include nearly every imaginable vegetable, and the number of possible combinations is wide-ranging. Perhaps best of all, they are very inexpensive.

A good soup can be highly nutritional and it can be the complete meal or only one course of a meal. Several types of soups can be made exclusively or nearly so from leftovers from the dinner table. In many instances, foods which are really not very palatable fresh can be turned into delicious soups, particularly if the food is somewhat beyond its peak ripeness or has reached the tough, hard-to-cook stage.

CANNING VEGETABLE AND STOCK SOUPS

Because nearly all types of soups can be canned easily, it would be virtually impossible to give a complete listing here. We can give you some typical recipes that work well, and we can provide a list of water-bath canning times and pressure-canning times. In general, however, you can time the processing in terms of the vegetable or meat that requires the greatest time and pounds of pressure.

Basic Soup Mixture

In so many soups you will see the same vegetables, and among the most common are tomatoes, peas, okra, corn, and string beans. The first basic soup mixture, therefore, will be a combination of these foods.

Start with 3 cups of finely chopped tomatoes that have vine-ripened but are still firm. Run the tomatoes through a food chopper. The corn on the cob can be dipped in boiling water and left to cook for 3 to 5 minutes before you cut it off the cob. Okra, too, needs to be blanched but only for 1 to 2 minutes, after which it should be chopped into 1/4-inch slices. You will need about 1 1/2 cups each of corn and okra, and 1 cup of peas which have been shelled and blanched for one minute. For extra taste you can add a small amount of chopped or grated onion and a teaspoon of salt and sugar, if you like. The sugar, incidentally, helps to neutralize the acidity in tomatoes, so you may want to include it.

Since the soup contains low-acid vegetables, these will need to be processed in a pressure canner rather than a water bath. When all the ingredients are ready, put the tomatoes to a boil and allow to simmer for 3 to 5 minutes, then add the other vegetables. Simmer the entire container full of vegetables for another 10 to 15 minutes and then pour into hot, sterilized jars and process at 10 pounds of pressure for the time needed for the longest-running vegetable. Since corn should be processed at 10 pounds of pressure for 85 minutes, this is the proper time to run the entire soup mixture.

When the processing has been completed, the soup can be eaten. The basic mixture here will probably be used only as a base for the finished soup, which may contain meat stock, leftover beef, or other meats. It can also be prepared as it is, thinned to taste with tap water, heated, and served. If you have leftover string beans, carrots, asparagus, or any other compatible vegetable, prepare it and add it to the soup mixture when you prepare it to serve.

Homemade Vegetable Soup

Start with 6 to 8 large tomatoes, which should be peeled, cored, and chopped finely. Then wash, peel, and dice 2 or 3 medium Irish potatoes. Grate or chop 2 medium onions, and add 2 washed and grated or sliced carrots. You can season by adding a lump of butter or margarine the size of a small hen's egg, 1 teaspoon of sugar, and salt and pepper to taste.

Cook the soup slowly for about 1 to 1 1/2 hours. If you wish, add 1 cup of corn, okra, squash, peas, beans, and any other vegetables you have left over from canning and let them simmer at least an hour. Pack into hot jars and process in the pressure canner at 10 pounds of pressure for 1 1/2 hours.

For a variation of the above recipe, cook a 3-pound soup bone with meat still on it until the meat is tender, and add the meat and stock to the mixture. Since beef should be processed for the same time as corn, you do not need to lengthen the processing time.

Meat-Stock Soup

If you wish to process only the meat stock to be used later with vegetable mixtures, or if you wish to enjoy it as it is, cover the meat bones and any meat scraps you have left over from a previous cooking with cold water and add a teaspoon of salt. Cook until the meat is completely done and tender. When the mixture starts to cool, the fat in the water will float and start to congeal, at which time it can be skimmed off with a large spoon. All bones and tendons should be removed before you process the stock. Any chunks of fat should also be removed. Pour the stock into hot, sterilized jars, close, and process at 10 pounds of pressure for 1/2 hour.

Chicken-Stock Soup

Cover chicken bones and meat with cold, salted water and cook until the meat is tender. Remove all fat, bones, and other waste and pack the stock into hot, sterilized jars. Process at 10 pounds of pressure for 1/2 hour. You can eat it as it is later, after thinning, if desired, with water and heating, or you can add it to a basic vegetable soup mixture.

You should, generally speaking, can stock from beef or chicken separately from vegetables and then combine the two when you are ready to serve. At serving time you can also add rice or noodles to the soup if you like.

CANNING INDIVIDUAL VEGETABLE SOUPS

You can make soups from individual vegetables easily, too, and serve these later. One example is asparagus soup, in which you can use asparagus stalks too tough to serve as a vegetable for the meal of the day.

Asparagus Soup

To make asparagus soup, cook the asparagus in just enough water to cover the stalks. When the asparagus, which should be cut into 1-inch sections to allow maximum cooking tenderness, is ready to process, press it through a colander or sieve and pack into hot, sterilized jars. You can add salt and pepper to your own taste and a small amount of grated onion or sweet peppers, if you wish. Process at 10 pounds of pressure for 40 minutes.

You can also mix the asparagus with meat stock canned separately and serve piping hot. You can thicken the soup with a spoonful of flour before you serve it.

Bean Soup

Any dried bean which has been cooked slowly over low heat will form a thick stock in the cooking liquid. If you cook 1 cup of dried navy beans, pinto beans, or white or great northern beans and add 1 quart of water, you can cook the beans for an hour or slightly longer. Beans that have not been presoaked will require longer cooking. Make sure to keep plenty of water in the pot.

When the beans are well done, press them through a sieve and return to the cooking liquid. Mix well and pack, still hot, into sterilized jars and process at 10 pounds of pressure for 50 minutes.

To presoak beans, put the beans in a pot, pour boiling water over them, and let stand until the water cools or for at least an hour. You can also put cold water over the beans and let them stand overnight.

For an even better bean soup, let the beans simmer over a woodstove heat or over an open fireplace, if you have one. If you cook the beans in a cast-iron pot with a heavy lid, when the steam from the pot pushes the lid up so that vapor can escape, you are in essence pressuring the beans. An added delight is the slight hint of smoke that permeates the beans as the lid is forced open for an instant. You can season the soup to taste when you serve it.

Cream of Celery Soup

Start with one cup of chopped celery stalks and leaves and cook until tender in three cups of water. In another pot cream three tablespoons of flour and equal amounts of margarine or butter, then add two cups of milk and heat until the liquid thickens. Push the celery through a sieve and add it to the liquid. Add salt and pepper to taste, and cook until all ingredients are mixed well. Pour into hot, sterilized jars and process at ten pounds of pressure for one hour.

Potato Soup

Peel, wash, and dice three large potatoes and two large onions. Put the mixture into one quart of boiling water. Cook until tender and season to taste before pouring into hot, sterilized jars. Process at 10 pounds of pressure for 40 minutes. When you serve you can add a pat of butter or margarine and serve hot.

You may also serve the soup by mixing it with other vegetable soups that you have preserved. Carrots, for instance, mix well with potatoes, and you can add tiny bits of freshly prepared beef for added flavor.

Processing

All vegetables listed here should be processed at ten pounds of pressure for the following times: asparagus, 40 minutes; lima beans, 50 minutes; string beans, 25 minutes; beets, 45 minutes; carrots, 30 minutes; corn, 85 minutes; mushrooms, 30 minutes; okra, 40 minutes; tomatoes, 40 minutes; peas, 40 minutes; potatoes, 40 minutes; succotash, 85 minutes.

Variations

You can make soups from many other vegetables. You can add onions, cream, flour, and seasoning before processing or as you serve the soup.

If you wish to make larger amounts of soups, simply multiply the measurements by whatever number is appropriate for your supply of vegetables or meat. Remember that you can add any extra vegetables and seasoning after you open the soup to prepare it for serving.

You can, in fact, use regularly processed

canned foods of all types for soups. The only drawback is that you may have to open a quart each of corn, tomatoes, beans, okra, and other vegetables, and the resulting soup will be far too much for one or 2 meals. You also don't always mix equal amounts of vegetables for the best soups, although you may certainly do so.

FREEZING SOUPS

You can freeze soup mixes easily by preparing the individual vegetables as described in the freezing section in Chapter 7 or, in the case of tomatoes, in Chapter 6.

To prepare large amounts of soup mix, you might want to cook 1 quart of dried beans, 1 quart of corn, 2 quarts of okra, 4 quarts of tomatoes, 1 quart of peas, 1 or 2 quarts of diced Irish potatoes, 1 quart or less of celery, 1 quart of asparagus, 1 quart of carrots, 1 quart of string beans, etc. When you have blanched and cooked all the vegetables for the prescribed times, pour the mixture into freezing cartons and store in the freezer.

As earlier stated, you can also freeze pints or quarts of the various vegetables and mix them when you prepare the soup. You can also freeze the meat or chicken stock separately to be thawed and mixed with the proper vegetables later. In all cases, prepare the foods you like best, regardless of recipes, and freeze amounts according to your own needs and appetites. If you live alone, then you should by all means freeze in pints.

DEHYDRATING SOUPS

You can dehydrate soups, just as you dehydrate other foods, with minor modifications. Instead of spreading the food to be dried on racks, you will need to make a container of aluminum foil, place the container on one of the dehydrator racks, and pour the soup into the container. Place the rack in the dehydrator and leave until all the water has been evaporated and all the vegetables or meats are free of normal moisture content. Then put the contents of the container into an air-tight container and store.

As the soup reaches the last stages of dehydrating, you will need to use a wooden spoon or similar utensil to make sure the food is not sticking to the aluminum foil.

A much simpler and less troublesome method is to cook the necessary foods for your soups, then spread them on the dehydrator rack and dry each item in a separate drawer or on a separate rack, since all soup foods do not require the same dehydrating time.

You can, for instance, cook and dry fragments or tiny cubes of carrots, potatoes, tomatoes, kernels of corn, okra, and all other soup ingredients, including the tiny pieces of meats that you would like to include. Store each ingredient in its own plastic bag, or all of the ingredients in one bag to be used later for soup.

If your foods do not require the same drying times, you can dehydrate one vegetable and store it, and, as others are ready, add them to the storage bag.

The advantage to this method is that when you are ready to prepare the soup, you can use only the amounts that you need. If you are eating alone, you can pour out enough to make only one bowl of soup, add the water to reconstitute, and heat as directed. If you have a dozen people dining with you, you can increase the amount of food accordingly.

The basic advantage to storing the foods separately is that if, on a certain occasion, you do not wish to have onions or celery, you may simply omit these items, or you may, if you prefer, add an extra generous serving of them.

SAUCES

To prepare sauces of all sorts for freezing, canning, or drying, you should prepare the food as directed in the chapters on preserving vegetables, fruits, or meats.

Spaghetti Sauce

An excellent spaghetti sauce can be made from the following recipe. Start with three teaspoons of shortening, margarine, or butter in a large pan or other cooking vessel. Chop three onions finely and add one large sweet green pepper, also chopped.

Brown the pepper and the onion in the shortening and add 2 pounds of ground beef. Brown all the ingredients together and then cover with 10 ounces of tomato puree, 3 teaspoons of tomato ketchup, 2 teaspoons of Worcestershire sauce, 2 teaspoons of vinegar, 1 teaspoon of hot sauce (or to suit your own taste), 1 teaspoon of paprika, 1/2 teaspoon of dry mustard, 2 teaspoons of chili powder, and 1/2 teaspoon of garlic powder. Simmer over low heat for 1 hour up to 2 hours.

When the sauce is completely done, you can pour it into freezer cartons. You can also put it into clean, hot jars and process for 75 minutes at 10 pounds of pressure. You can make an aluminum foil basket and dehydrate the sauce and then store it in air-tight containers.

Freezer cartons should be kept out of the freezer until the sauce has cooled in order to keep from frosting your freezer and causing ice crystals to appear in the sauce.

Other Sauces

You can prepare other sauces from a wide range of recipes and to meet your own tastes. Most sauces are better when used fresh, so we do not recommend preserving large amounts of sauces. We suggest instead that you preserve the ingredients separately so that you can make your sauces fresh. If you do preserve them, we suggest freezing as the best method.

10

Making Jams, Jellies, and Preserves

Among the most delicious and colorful foods in your kitchen are jams, jellies, preserves, conserves, and other fruit preparations. A hot biscuit, a pat of butter or margarine, and a spoonful of homemade jelly can be, in fact, a complete breakfast that is tasty, filling, and nourishing (Fig. 10-1).

Homemade jams and jellies bear little if any resemblance to those jars of colored gelatin flavored with bits of fruit found on supermarket shelves. It is unlikely that anyone who has eaten the real thing can ever return to a diet of mass-produced types fruit preserves.

If superb taste is not enough, the cost of making your own jellies and jams is only a fraction of the cost of a jar of supermarket preserves. A jar of preserves will cost about $2 in the store, but you can make the same amount for much less. For the cost of 32 ounces of strawberry jam, for instance, you can make about 154 ounces of homemade jams.

The process is simple and reasonably easy Among the best persuasions for making the jellies is the fact that you can use materials that you might ordinarily discard. Apple cores and peelings, for instance, make wonderful apple jelly. Grapes that have been squeezed for juice can be cooked to extract additional juice for jellies. Corn cobs, as strange as it may seem, can be cooked and the juice made into jelly. You are also using the natural nutrients rather than partaking, at breakfast, of a jelly that is nearly all granulated sugar and few if any nutrients.

CANNING RECIPES

Nearly every fruit and berry can be used for some type of jelly or preserve. Each recipe is so different in at least one important facet that numerous recipes will be given in this chapter.

Using Apples

If you have canned, dried, or frozen apples, save the cores and peelings to make fine jelly from them.

Apple Jelly. You can also select sound, firm apples, wash, but do not peel. Slice them into strips 1 inch wide at the center and trim away the seed

Fig. 10-1. Jams, jellies, and preserves made from fresh fruits and berries are among the most delectable items you can place on the breakfast table. The cost of making these delicacies is minimal.

and pulp areas. Put the slices, with the peel still on them, into a pot and cover with water. Cook until the fruit is soft.

Press the fruit through a jelly bag or cheesecloth, or anything that will hold the pulp and allow the juice to seep through (Figs. 10-2 and 10-3). Strain the juice extracted to remove all solids and then add 3/4 cups of sugar to each cup of juice, but make sure the juice is boiling before you add the sugar. Cook until the jelly is ready.

You can test for readiness by tilting a spoonful of jelly sideways. When the jelly no longer runs off in a steady stream but in 2 drops at a time, it is ready. Pour the jelly into sterilized glasses and allow it to set. Jelly should be able to hold its shape when spooned onto a plate, and it should be clear and bright.

To seal the jars, wait until the jelly is cold, then clean the rim of the jar with a damp cloth. Melt paraffin over low heat. Pour the melted paraffin into the top of the jelly jar and allow it to cool. The wax will seal the top of the jar effectively. Cover the wax with a clean lid and store the jelly until you are ready to use it.

If you don't want to use paraffin, you can pour the hot jelly into hot, sterilized jars, leaving 1/8 inch of space at the top. After wiping the jar rim clean, you can place a metal lid on the container and screw the ring down firmly. For best results lids should be placed in boiling or very hot water before using. Be sure the sealing compound on the lid is properly positioned when you tighten the rings. The jars will then vacuum seal and can be stored at room temperature for months.

Do not freeze the jelly; however, you may freeze the apples and later use them for making preserves.

Apple Butter. Core and slice apples but do not

Fig. 10-2. To make a jelly bag, sew a hem into a piece of cloth and allow enough room for a large stick to pass through. A broomstick is the right size. Now sew the bag closed, except for an opening at the top where fruit pulp will be inserted. Sew another hem into the bottom and allow enough room to pass the handle of a long wooden spoon through it. You can rest the broomstick across the backs of chairs and place a vessel beneath the bag to catch the juice that is squeezed out.

Fig. 10-3. While the broomstick holds the bag in place, twist the wooden spoon until the juice is squeezed through the bag and into the container below. One person can operate the simple jelly bag effectively.

peel. Place them in a pot and cover with water. Cook until apples are soft, then press through a sieve. Measure out 16 cups of apple pulp and add 1 cup of vinegar and 8 cups of sugar. You may also add, if you desire, 2 to 4 teaspoons of cinnamon.

Cook all ingredients until it turns into a smooth, even consistency, usually about 1 1/2 hours. Pour into hot, sterilized jars, seal with a hot lid, and tighten the ring firmly. You can also freeze the apple butter in plastic cartons or process for 5 minutes in a water-bath canner.

Applesauce. Wash, pare, and core apples and then simmer in a pot with enough water to cover the bottom of the pot. When apples are tender, press them through a sieve and sweeten to taste. Many people add 1/4 cup of sugar for every 4 medium apples, but we have found that if we use about half the prescribed amount the apples have a more natural taste. You do not need to sweeten at all if you enjoy slightly tart apples. When apples have cooked fully, pour into jars, leaving 1/8 inch of space, and close the jars. Process 20 minutes in a water-bath processor.

Apple Marmalade. Use 8 pounds of apples that have been washed, peeled, cored, and sliced. Cook until the apples are a soft pulp, then add 4 pounds of sugar and 15 ounces of crushed pineapple. You can buy commercially canned pineapple or, better still, buy a whole pineapple and crush the meat yourself. Process for 10 minutes in a water-bath canner. If you don't have pineapple, you can substitute crushed orange pulp. In all instances you will want to use citrus fruits in marmalades.

Combination Jellies. You may combine other juices with apple juice to make combination jellies. You can mix equal parts of blackberry juice, elderberry juice, or other juices that appeal to you when paired with apple juice. For each cup of mixed juice use 1 cup of sugar and proceed as you did with apple jelly.

Apple-Peach Conserve. Cook 2 cups of chopped and cored but unpeeled apples, 2 cups of chopped and peeled peaches, 1/2 cup of lemon juice, 3 cups of sugar, and if you desire, 1/2 cups of nuts and/or raisins. When all ingredients are fully cooked and the apples are transparent, pour into hot, sterilized jars or glasses. You can seal them hot or you can process in a water bath for 10 minutes.

Using Apricots

You can make a variety of preserves with apricots, as well.

Apricot Jam. If you have fresh, ripe apricots, wash and pit, then remove skins. If you have dried apricots, simmer them in warm water until they are reconstituted. Crush apricots and add 2 cups of sugar for each cup of crushed apricots. Add 1/2 sliced lemon, and, if you prefer, 2 cups of crushed pineapple. If you add the pineapple, add another 2 cups of sugar. The warm water you simmered the apricots in—2 1/2 cups—should be retained in the cooking of all the blended ingredients. Continue cooking until the jam is very thick and nearly clear. Pour into hot, sterilized glasses or jars and seal with paraffin or process in a water bath for 5 to 10 minutes.

Apricot Butter. Wash pit, and chop apricots, add a cup of water and cook until the apricots are soft and pulpy. Press through a sieve or cheesecloth and measure the amount of pulp you have. For each cup of pulp add 2/3 cup of sugar and either 3 tablespoons of lemon juice or 4 tablespoons of orange juice. Cook all ingredients together until a uniform consistency is reached. Pour into hot, sterilized jars and process for 10 minutes in a water-bath canner.

Apricot Conserve. Wash, pit, and chop about 4 pounds of apricots, add 5 chopped, seeded and peeled oranges, 5 cups of sugar, 1 tablespoon of lemon juice, and 1 cup of raisins or chopped nuts. Do not add the nuts until the cooking is nearly completed—the last 4 minutes or so. Boil until all ingredients are cooked together and the mixture thickens well. Then pour into hot, sterilized jars, close, and process for 10 minutes in a water bath. You may also substitute crushed pineapple, for oranges, if you prefer.

Using Blackberries

Blackberries, and several other berries, such as

dewberries, raspberries, and others, can be made into wonderful preserves.

Blackberry Jam. Crush and cook the berries in 1/4 cup of water for each 4 cups of berries and 2/3 to 3/4 cup of sugar for each cup of crushed berries. Cook until a mild consistency is reached, and then press through a sieve to remove seeds or leave the seeds in and pour the jam into hot, sterilized jars. Seal with paraffin or process in a water bath for 10 minutes.

Blackberry Jelly. Place washed blackberries in a cooking vessel and slowly bring to a boil. While cooking the berries, crush them slightly to aid in bringing out the juice. Do not add water or sugar at this step (Fig. 10-4). After cooking and crushing the berries for about 15 minutes, strain the contents through a jelly bag. Measure the juice and return it to a cooking vessel to bring to a boil. For each cup of juice add 3/4 cup of sugar. Do not make batches larger than 4 cups of juice and 3 cups of sugar, for best results. Add the sugar gradually to the boiling juice. Continue cooking the juice and sugar for 15 minutes or until jelly drops from a spoon, 2 drops at a time. A jelly thermometer can be used instead. Fill jelly jars to within 1/4 inch of top and seal, or allow jelly to cool and firm before covering it with a layer of paraffin (Fig. 10-5).

Using Cherries

Cherries make luscious preserves.

Cherry Preserves. Place 4 cups of pitted and washed cherries in a large saucepan and mix well with 2 cups of sugar and 1/2 cup of water. Bring contents to a boil and continue cooking for 10 minutes. Then add 2 more cups of sugar and boil for another 10 minutes. Shake the pan while the

Fig. 10-4. Cook the berries and crush lightly to bring out the juice. Strain the juice through a jelly bag later.

Fig. 10-5. If you wish you can modify the jelly recipe slightly and make delicious pancake syrup. Add sugar to the blackberry or other juice. Instead of cooking the juice and sugar until it thickens, cook only slightly to cause a gentle thickening. The juice should be the consistency of thick pea soup.

boiling is taking place to prevent the cherries from sticking and burning. After cooking, remove the pan and let the contents stand overnight to prevent fruit from floating in the jars. Pour cold into sterilized jars and seal with paraffin or process the closed jars in a water-bath canner for 20 minutes.

Cherry Conserve. A tasty cherry conserve can be made from 3 cups of washed and pitted cherries, 1 cup of raisins cut in pieces, and 3 cups of water. Bring to a boil and continue cooking for 30 minutes. Then measure the pulp and juice, and for each cup, add 1 cup of sugar. Cook the sweetened mixture rapidly until thick, stirring constantly. The sugar must be completely dissolved. Pour into sterilized jars and seal.

Cherry-Pineapple Conserve. A cherry-pineapple conserve can be made with 4 cups of pitted and chopped cherries and fifteen ounces crushed, drained pineapple. Add 3 cups of sugar to the cherries and pineapple and boil the mixture until thick. Stir occasionally to prevent scorching. Before pouring the cooked mixture into hot jars, add 1/2 cup of nuts, if desired. Seal.

Using Cranberries

Cranberries can also be used in various preserves.

Cranberry Jelly. To 4 cups cranberries add 1 cup of water and cook until cranberries are tender. Strain the contents. To the juice add 2 cups of sugar and boil for 3 minutes. Pour into jars and cool, if sealing with paraffin, or process in a water bath for 10 minutes.

Cranberry Jam. Crush lightly 8 cups of cranberries. Add 1 cup of water and 1 cup of vinegar. Cook until the cranberries are tender and strain through a colander. Anything that serves as a coarse strainer will do. After the straining add 6 cups of sugar, 1 teaspoon of cinnamon, 1/4 teaspoon of cloves, and 1/2 teaspoon of allspice. Cook the contents 8 minutes, stirring frequently. Pour into sterilized jars and process in a water-bath canner for 10 minutes. The spices can be measured more or less to suit your tastes.

Fig Jam

Use about 3 quarts of sound and firm figs. Wash and pour boiling water over them. Let the figs stand for 1/4 hour in the hot water, rinse, and drain. Make a syrup of 4 cups of sugar and 3 quarts of boiling water. Drop the figs slowly into the water and cook until they are transparent. Remove the figs, continue cooking the syrup until it starts to thicken, and pour it over the figs. Let the mixture stand for 8 to 10 hours and then put the figs and syrup into hot, sterilized jars. Process in a water bath for about 1/2 hour.

Using Grapes

Grapes can be used in several ways to make delicious spreads.

Grape Jelly. Use 4 pounds of grapes. Wash, crush, and cook on high until the 1 pint of water you poured over the grapes has started to boil. Reduce heat and simmer until the fruit is soft. Press through cheesecloth and measure the grape juice. Add 3/4 cup of sugar to every cup of juice. Mix sugar and juice and cook until mixture thickens to the double-drop thickness. Remove from heat and pour into hot, sterilized jars. Process for 10 minutes in a water-bath canner.

Grape Preserves. Place 4 pounds of grapes in a pot and cover with water. Cook until the grapes, which you should crush as they cook, are tender. Remove from heat and press the pulp through a sieve. All seeds and hulls should be held out by the sieve. Add 1 cup of sugar for every 2 cups of grape pulp you have. If you need greater thickness, add 1 1/2 cups of sugar for each 2 cups of pulp. Cook until desired consistency is attained. Remove from heat and pour into hot, sterilized containers and close. Process for 10 minutes in a water bath.

Grape Butter. Wash grapes, crush, and remove hulls. Cook until the pulp is soft and tender. Run through a sieve to remove seeds. Return the pulp to the cooking vessel and add 1 cup of water for each pound of pulp—or 1 cup per pint—and add 2 1/4 cups of sugar per pint of pulp. Cook until thick, then pour into hot, sterilized containers and process for 10 minutes in a water bath.

Grape Jam. Start with 6 pounds of grapes. Wash, stem, and remove the hulls. Place the pulp in a cooking vessel, add 2 cups of water, and bring to a boil. Cook until pulp is tender, then remove from heat and put the pulp through a sieve to remove seeds. Return the pulp to the burner and add 12 cups of sugar and 2 boxes of fruit pectin slowly as the mixture returns to a rolling boil. Continue cooking until the jam thickens, then remove from heat and pour into hot, sterilized jars. Close and process in a water-bath canner for about 10 minutes.

Using Peaches

Your family is sure to enjoy the various forms of preserving peaches.

Peach Jelly. Peel and pit peaches, then cover with water and boil until the juice is extracted. Remove from heat and allow to cool, then press through a sieve. Measure the amount of juice you have and add 3/4 cup of sugar to each cup of juice. For each pint of juice add 1 pint of fruit pectin or 1 box of powdered fruit pectin. Continue to cook until the jelly thickens and double-drips from the spoon, then pour into hot, sterilized containers and close. You can process for 10 minutes in a water bath or seal with melted paraffin.

Peach Jam. Use 2 to 2 1/2 pounds of peaches, which you should wash, peel, pit, and chop finely. Add to the fruit a box of powdered fruit pectin and 4 cups of sugar. Put the vessel on the burner and heat until the fruit mixture is boiling. Reduce heat and simmer until the jam thickens, then pour into hot, sterilized containers and close. You can process for 10 minutes in a water bath, or you can seal with paraffin.

Peach Butter. Wash, peel, and pit peaches, then cook the pulp until it is tender. Put through a sieve. Add 1/2 cup of sugar to each cup of pulp. Then cook until the mixture is thick and clear. Pour into hot, sterilized jars and seal with paraffin or process for 10 minutes in a water-bath canner.

Using Pears

Do not overlook the various uses of pears when making preserves.

Pear Honey. Wash, peel, and remove the cores from about 6 pounds of fresh, firm pears. Chop finely and then feed through a food chopper. Add 2 cups of crushed pineapple and the juice of 2 limes and grated lime rind. Put in 10 cups of sugar, then cook the mixture until it thickens. Pour into hot, sterilized containers. Process for 10 minutes in a water bath or seal with paraffin. Like most preserves, it will keep for several weeks in the refrigerator.

Pear Preserves. Choose firm, fresh pears, and peel, seed, and chop 4 pounds. Cook in 4 cups of water until they are easily pierced with a fork or until they can be crushed easily. Remove the pears and make a syrup of the cooking water and 7 cups of sugar. Bring the syrup to a boil. Return the pears to the syrup and cool until the pears have a clear and thick appearance. Remove from heat and pour into hot, sterilized jars and seal with paraffin or process in a water bath for 10 minutes.

Using Plums

Plums, too, make wonderful preserves.

Plum Jam. Wash the plums and then boil until the skins are easily pierced with a fork. Crush and press through a sieve, then return the juice and pulp to the cooking vessel and add 1 cup of sugar for each cup of pulp. Cook until the mixture is thick and put into hot, sterilized jars. You can seal with paraffin or process for 10 minutes in a water bath.

Plum Preserves. Start with 2 quarts of pitted plums, and pour 6 cups of sugar over the chopped plums. Add 1 cup of water and cook for 20 minutes or until the plums are tender. Reduce heat and simmer until the syrup and plums start to thicken. Remove from heat and seal with paraffin or process 10 minutes in a water-bath canner.

Using Strawberries

Of course, strawberry preserves are favorites of many people.

Strawberry Jelly. Wash and cap strawberries, then crush and put 1 cup of sugar over 1 cup of strawberry juice after the pulp has been put through a sieve and only juice remains. Add 1 cup of fruit pectin and cook all the ingredients until the jelly is of proper consistency. When it is ready it will run off the spoon in 2 droplets. Take it from the heat and pour into hot, sterilized jars. Seal with paraffin or close with lids and rings and process for 5 minutes in a water-bath canner.

Strawberry Preserves. Wash, cap, and drain berries. Measure out 1 quart of berries, put into cooking vessel, and add 1/3 cup of water and 3 cups of sugar. Cook until sugar is dissolved and berries can be crushed easily. Stir the berries in the mixture frequently, and, when the berries and syrup are thickened, pour into hot, sterilized jars or glasses. Seal with melted paraffin or process in a water bath for 5 minutes. After you have sealed the jars, allow them to cool, then turn them upside down and leave for 1/2 hour, then return to normal position. If the berries tend to sink to the bottom, turn the jars upside down again and leave for another 1/2 hour.

Tomato Preserves

Wash 5 pounds of tomatoes and then dip into boiling water until the skins can be removed easily. Remove from water, slip peels off, and slice or quarter the tomatoes. Add 7 cups of sugar and allow the tomatoes to stand overnight or for 8 hours. Then drain and boil the syrup and sugar mixture until it starts to thicken. Add the tomatoes and the juice of 2 lemons. When this mixture thickens, remove from heat and pour into hot, sterilized jars. Seal with paraffin or process in a water-bath canner for 10 minutes.

Watermelon Rind Preserves

After you eat the melon meat, save the rinds and trim off all the outer peel and all the meat until you have only the white rind left. Cut the rind into thin strips 1 inch long and 1/2 inch wide and thick. Mix 1/2 cup of salt to 1 gallon of cool water and soak the rind in this mixture for 8 to 10 hours. Then drain, rinse, and cook in water until the rind is clear or transparent. Drain again and make a syrup of 8 cups of sugar and 8 cups of water, the juice of 4 lemons, and any spices, such as cloves or cinnamon, that you desire. If you use spices, also use a spice bag so you can remove the spices after cooking.

Boil the syrup 5 to 7 minutes. Then add the watermelon rinds and cook until perfectly well done and transparent and tender. When the mixture has thickened to the desired point, remove from heat, remove the spice bag, and pour the preserves into hot, sterilized jars. Close the jars with rings and lids left in very hot water until used and process in a water-bath canner for 10 minutes.

There are, of course, many other recipes and many other fruits that can be used for jams, jellies, and preserves. You can also use your own imagination and taste and combine various fruits with other citrus fruits or non-citrus fruits. Apples and strawberries, for example, blend well, as do peaches and pineapple, peaches and pears, peaches and plums, and pears and raisins.

OTHER WAYS TO PRESERVE

Some jams and jellies can be frozen, but most require more sugar than we enjoy. In fact, when you are making any of these jams or preserves, feel free to cut back on the sugar as much as you can, unless you enjoy them extremely sweet. You need a certain amount of sugar, however, for the fruits to jell or set, but you can use fruit pectin to help the setting, or you can use commercial products.

Instead of freezing the jams and jellies we recommend that you can, dry, or freeze the fruits and berries. When you want jams or preserves, open a quart, thaw a carton, or add water to rehydrate the dried foods and proceed as the instructions direct.

11

Preserving Desserts

One of the best ways to make sure you will not run out of dessert materials in mid-winter is to preserve your desserts, ready to eat, or to can or freeze the fillings, ready to eat except for crusts, when the fruits and berries are plentiful.

The only two drawbacks, in addition to the cost of running the freezer or canner, are that you must spend a great deal of money on sugar and other necessities, and you must have your freezer and canning space occupied for a considerable time.

The advantages far outweigh the disadvantages. It is not likely that you will be able to buy sugar during the winter any cheaper than you can in the summer. In fact, as inflation steadily gnaws away at our budget, the cost of all foods seems to climb steadily; so you might save a considerable amount of money by canning or freezing your dessert materials in the summer, particularly if you buy in large quantities, which are often cheaper in the long run.

As for the freezer, it will probably be running anyhow, since there will be other food in it, so adding desserts will not increase the power bill appreciably. The cost of canning is, no matter how we try to rationalize it, an additional expense. You can, however, freeze or can desserts or dessert ingredients much cheaper than you can purchase them in the grocery store.

The basics of freezing and canning dessert foods are simple. Once you have mixed the ingredients, following whatever recipes you enjoy most, simply process in the water-bath canner or pressure canner for the required amount of time. If you freeze the desserts, after the recipe is mixed wrap the dessert in aluminum foil or plastic freezer bags and store them in the freezer so that they will not be damaged by other foods and so you can get them out easily when you need them.

Nearly any kind of cake, custard, cobbler, ice cream, or sherbet can be frozen very easily. Frozen pies can be stored as fillings or as cooked or uncooked pies. As for canning, you can preserve the basic pie or cobbler fillings. When you are ready to use the fillings, simply prepare the crust, pour the filling into the crust, and bake the pie.

CANNING PIE FRUITS AND FILLINGS

You can preserve the dessert materials according to the instructions given earlier and can only the fruits and berries. This allows you a great deal of freedom of choice later in the way that you will use your fruits and berries.

If you elect to can the pie fillings, we suggest that you use quart rather than pint jars if you find it handy to bake 2 pies at a time. Of course if you have a very small family or one that eats very few desserts, you may find it better to bake one pie at a time.

On the other hand, you can always freeze the second pie and thaw it for consumption later. If you are going to be operating your oven, you might as well make two desserts for the power cost of one.

These recipes will provide a general direction for your own canning. Do not include flour, eggs, margarine, and similar products in your canned filling. You can add these easily when you are ready to bake the pie.

Apple Cobbler

Slice peeled and cored apples thinly and measure. You need 2 full cups packed well. Add 3/4 cup of granulated or brown sugar and 1/2 teaspoon of cinnamon. Put 1/4 cup of water in the bottom of a cooking vessel, add sliced apples and other ingredients, and cook until tender. You may add other spices, if you like, and raisins. Process the way you would canned apples.

When you open the jars to make cobblers, prepare your crust and pour the filling into the crust, which is in the baking pan. Over the surface of the cobbler fruit 1/3 of a stick of margarine or butter should be dotted so that during baking the butter will melt and spread evenly throughout the cobbler. Add 3 tablespoons of milk and 2 tablespoons of flour to act as a thickener. Cover the fruit with the top crust and bake at 350 degrees for 30 to 40 minutes.

If you want a larger cobbler, simply double the ingredients and bake as usual.

Apple Pie

Peel, core, and slice 6 apples thinly and place in cooking vessel. Add 1/2 cup of water and 1 cup of sugar, plus 1 teaspoon of cinnamon. Add 1 teaspoon of lemon juice or the juice and pulp from 1/3 of a lemon. Add 1/2 cup of raisins which have had boiling water poured over them and have stood for 15 minutes to plump well. Cook the mixture until all ingredients are mixed well and the apples are very tender. Process as you would canned apples.

Before you add the top crust mix 1 tablespoon of flour with 2 tablespoons of milk and distribute over the pie surface, then dot with butter or margarine. Bake for 35 minutes at 350 degrees.

Blackberry Cobbler

For every 3 1/2 cups blackberries add 2 cups of granulated sugar and 1/2 cup of water. Crush the berries slightly (Fig. 11-1) and cook until they are very well done and the sugar is completely dissolved. Process as you would blackberries for canning.

When you start to bake the cobbler, if your berries are extremely juicy you may need to add 1 or 2 tablespoons of flour to act as thickener. Pour the mixture into a large baking dish that has a bottom crust already prepared, then dot the surface with butter or margarine, add the top crust, and bake for 40 minutes at 350 degrees. Serve with whipped topping or ice cream or eat it as it is. You may also wish to serve piping hot cobbler with cream poured over it.

If you don't want to prepare a crust, you can pour the filling into a pot and heat until the boiling point is reached, then drop strips of uncooked dough into the filling or small balls of dough the size of a quarter and let the dough cook and expand in the filling (Fig. 11-2).

Blueberry Pie

Mix 2 1/2 cups of blueberries with 1 cup of granulated sugar, a dash of salt, and 1 large slice of lemon which has had the juice and the pulp squeezed out. Discard the peel and add the juice and pulp to the recipe. Can as you would process blueberries.

When you are ready to use the filling, prepare your pastry and pour the filling into the bottom

Fig. 11-1. When the berries are in the bowl, add sugar and stir until the berries are coated well and the juice begins to seep from them.

Fig. 11-2. When the cobbler is complete, it can be frozen as it is or you can bake it first and then freeze the fully cooked dessert. To use the cobbler, you can bake the uncooked one while it is still frozen. You can thaw the cooked one by placing it in a slightly warm oven for an hour before meal time.

crust, then sprinkle 2 or 3—depending upon the amount of juice in the berries—tablespoons of flour over the filling. Dot with butter or margarine before adding the top crust. Bake at 400 degrees for 40 to 50 minutes.

Cherry Pie

To make this pie, or cobbler, if you prefer, use 2 cups of pitted, stemmed, and washed cherries, 1 full cup of sugar, and 1/4 cup of water. Cook the ingredients until the cherries are tender and all the sugar has been dissolved. Process as you would canned cherries.

When you prepare to bake, add 1/2 teaspoon of almond extract and 1/4 cup of flour to act as filler. Dot with butter, put on second crust, and bake 40 to 50 minutes at 400 degrees.

Peach Pie

Peel, pit, and slice peaches until you have 6 cups. Add 3/4 cup of sugar, 1 tablespoon of lemon juice, and 1/2 teaspoon of ground cinnamon. Place all of the ingredients in a cooking vessel and heat to boiling. Cook until peaches are soft and tender and all the sugar has been dissolved. Process as you would can peaches.

When you are ready to bake the pie, spread pastry in a 9-inch pie pan, pour in the filling, and add 2 tablespoons of flour—3 if the peaches are very juicy—and dot with butter.

If you prefer, you can substitute 2 1/2 tablespoons of cornstarch and add 1 tablespoon of corn oil margarine instead of butter. Bake for 40 minutes at 425 degrees.

Peach Cobbler

Peel, pit, and slice enough peaches for 4 well-filled cups, and place in a cooking vessel. Add 1 cup of sugar, 1/4 cup of water, and 1/8 teaspoon of almond extract. Bring to a boil and cook until the peaches are tender and the sugar is dissolved. Process as you would canned peaches.

When you are ready to bake, pour the filling into a deep dish pastry shell. Add 1/3 cup of brown sugar and 1 1/2 tablespoons of cornstarch or flour for thickener and dot 3 or 4 tablespoons of butter over the top. Add the top crust and bake for 40 to 50 minutes at 400 degrees.

Wild Strawberry Pie

Wild strawberries outdistance virtually every other type of fruit or berry in the most distinctive and delicious flavor imaginable. The berries are small; so it takes a long time to pick enough for a pie, but when you have enough for 1 cup of crushed berries, you can wash and cap the berries and place into a cooker. Add 1 cup of sugar, 1 teaspoon of lemon juice, and cook until berries are tender and the sugar is dissolved. Can as you would process strawberries.

When you bake the pie, add 3 tablespoons cornstarch or flour and spread the filling over the pastry crust. Dot with butter and bake for 40 minutes at 400 degrees. Serve with cream or whipped cream topping.

Rhubarb Pie

Prepare rhubarb as you would for canning, and add 1 1/2 cups of sugar for every 3 cups of chopped rhubarb. Also add proportionately 1/2 of an orange that has been seeded. Include the pulp as well as the juice. Discard the peel and add 1/2 cup of brown sugar. Can as you would process rhubarb.

When you bake, place the filling in the shell and sprinkle about 3 tablespoons of flour over the mixture and dot with butter. Bake 40 minutes in a moderate oven.

FREEZING DESSERTS

Canning dessert filling is at times fairly limited, but freezing desserts seems to know no limits. You can freeze the desserts uncooked, baked, or in stages (Fig. 11-3).

If you happen to have an abundance of eggs and other ingredients for cakes, bake just the layers and then later, after you have frozen the layers, complete the cake with whatever filling or frostings you prefer.

Layer Cake

A simple and very satisfactory layer cake can be

Fig. 11-3. You can freeze fruits, pie fillings, or the entire dessert. One good practice is to make two cobblers when you are baking so that you can enjoy one hot and then freeze the other for later use.

made by starting with 2 cups of enriched flour, and adding 1 1/3 cups of sugar, 2 or 3 eggs, 1 cup of milk or slightly less, depending on the amount of moisture in the eggs, 1 1/3 sticks of margarine or butter, 2 teaspoons of baking powder, and 2 teaspoons of vanilla flavoring or 1 teaspoon of vanilla and 1 of lemon flavoring.

It is commonly stated—and believed—that baking a cake can be accomplished only by doing everything exactly right, such as sifting flour, adding only the dry ingredients first, etc. We put every ingredient into the mixing bowl at the same time, and we do not necessarily use softened butter or margarine. We simply mix until the batter is smooth.

Bake this cake for 30 minutes at 350 degrees, but be sure to pour the batter into greased and floured cake pans. When the baking is completed, remove the cake from the oven. As soon as you can, remove the layers and place them on a towel to cool. Do not cover the layers. When they are cool, wrap them in aluminum foil or place them in large plastic freezer bags and put them in your freezer.

If the batter looks too thin, add more flour; if it is too thick, add more milk. We have never had the slightest hint of a problem, let alone a failure. It is simply a foolproof layer cake. Sometimes if we plan to use egg whites for a 7-minute frosting, we put only 1 egg and the yolks of the eggs used in the frosting into the batter.

It is a good idea to bake several of these layers while you have spare time and freeze them. When you need a quick dessert, all you need to do is pull out two or three of the layers, add frosting, and serve. Thawing takes less than 1/2 hour, and your layers can easily reach serving temperature while you are preparing the rest of the meal. You can also place the layers in an oven that has been recently heated to bake bread and which still holds enough heat to thaw the cake, but not enough heat to cause drying out or burning.

If you wish, you can frost the cake before freezing it. We have frozen boiled frostings, cooked frosting of all sorts, and, despite the warnings that such icings do not withstand freezing, we have never had any trouble. If you are reluctant to try cooked frostings, you can make a confectioner's sugar frosting, and it will freeze very well.

Seven-Minute Frosting

After the layers are baked, make a 7-minute frosting by combining 2 egg whites, 1 1/2 cups of sugar, 1 1/2 teaspoons of corn syrup or 1/4 teaspoon of cream of tartar, and 1/3 cup of water, and a pinch of salt. The salt is optional. Mix the ingredients well for about 1 minute by beating. Place the pot inside another pot containing boiling water—a double-boiler is perfect—and continue beating until peaks form and hold in the frosting. While this is called a 7-minute frosting, it usually is ready in 4 or 5 minutes.

Spread the frosting on the cake and, for finishing touches, cover each layer with frosting and a liberal spreading of shredded or grated coconut. When 1 layer is completed, stack the next layer on top, frost it, and cover with coconut. Then frost the sides and cover with coconut.

You can freeze the entire cake as it is, and when you need to serve it, remove it from the freezer an hour or two before mealtime.

Confectioner's Frosting

You can also make a frosting of confectioner's sugar. Start with 2 cups of confectioner's sugar and add enough milk, cream, or margarine to mix into a good spreading consistency. After the mixing is completed, you can add a teaspoon of vanilla and mix well again. You can also add the vanilla flavoring to the 7-minute frosting described above. After the frosting has been removed from the heat, beat it while adding the vanilla, and continue to beat until the vanilla has been completely mixed.

Variations

For a delicious variation on the above cake, bake the layers, then peel, pit, and slice well-ripened peaches and cover the layers with icing, then with sliced peaches, and allow to set for 3 hours before sampling. This cake does not freeze well, but once you sample it there won't be enough left to freeze anyway.

Another variation is to squeeze five or six oranges until the pulp is also removed from the oranges and is mixed with the juice. Use a fork to perforate the cake layers repeatedly, and then spoon the orange juice and pulp onto the layers. The juice will tend to be absorbed by the layers and the pulp will remain on top. Cover with 7-minute frosting and freeze the cake by placing it in a large plastic freezing bag or by wrapping it in aluminum foil.

If you like bananas, you can spread frosting over the layers and then slice several bananas thinly. Spread the slices over the cake. Do this with both layers, but do not freeze this cake since the bananas will turn dark in the freezer. It is excellent fresh or with layers that have been removed from the freezer.

Apple-Nut Cake

One cake that freezes beautifully is an apple-nut cake that is prepared in the following fashion. Chop or grate 3 cups of peeled apples. Set aside. In a large mixing bowl combine 1 cup of cooking oil—2 tablespoons less if apples are really juicy—and 2 cups of sugar. Add 2 eggs and 1 1/2 teaspoons of vanilla. Sift together 3 cups of all-purpose flour, 1 teaspoon of soda, 1/2 teaspoon of salt, and add to the sugar-oil-egg mixture. Stir in chopped apples and 1 cup of chopped nuts. You may have to stir by hand at this point. Thoroughly combine the ingredients.

Cook the cake in a greased and floured tube pan which also has a waxed paper liner in the bottom. The cake can also be baked in a 9 x 13 loaf pan. Bake for 1 hour or until done at 350 degrees.

While the cake is cooling, make the following caramel tasting glaze. In a double boiler melt 1/2 cup of butter or margarine and 1 cup of brown sugar. The brown sugar requires 5 minutes usually to lose its graininess. Then add 1/4 cup of cream or evaporated milk and bring the mixture to a full boil. Remove from heat and cool slightly before adding 3/4 teaspoon of vanilla. Generously cover the cake. If you wish to drizzle the cake lightly, use only half the amounts given and you will still have an ample supply of glaze.

When the cake and glaze have completely cooled, freeze the cake first and then wrap it. If you baked the cake in a tube pan, you should place some freezer wrap in the hole in the center to make sure the cake is properly sealed.

Strawberry Cake

A delicious cake features strawberries in the batter as well as in the icing. You may use a bought white cake mix, 3 teaspoons of plain flour, 1 box of strawberry gelatin, 3/4 cup of drained strawberries—reserve the juice—1/2 cup of water, and 3/4 cup plus 2 tablespoons of cooking oil. Combine the above ingredients and add 4 eggs. Bake at 325 degrees for 35 minutes.

If you desire a cake made from scratch, however, combine 3/4 cup of margarine, 1 1/2 cups of sugar, 3 eggs, 3/4 cup of milk, 2 cups of all-purpose flour, 2 teaspoons of baking powder, 1 small box of strawberry gelatin, 3/4 cup of strawberries, drained, with juice reserved. Pour into greased and floured pans. Bake for 25 to 30 minutes at 350 degrees.

When the cake layers have cooled, cover them with the following icing. Use the strawberry juice and any leftover strawberries to make 1/2 cup. If necessary add water to reach this measurement. Combine the strawberries and juice with 1 stick of softened margarine and 1 box of confectioner's sugar. Add a pinch of salt. This amount will generously ice two layers or a large sheet cake.

When the cake is completely cooled and iced, wrap it in freezer wrap or heavy foil and freeze.

Best-Ever Pound Cake

A pound cake that freezes excellently is the Best-Ever pound cake. Start by creaming 3 cups of sugar, 2 sticks of margarine or butter, and 1/2 cup of shortening. Add 6 eggs, 1 at a time. If you prefer a cake with less cholesterol, use 4 eggs rather than 6. Beat the batter well, and add 3 1/4 cups of enriched flour. Add slowly, then sprinkle into the batter 1/2 teaspoon of salt and 1/2 teaspoon of baking powder, 1 teaspoon of vanilla extract, and 1 teaspoon of lemon flavoring. As you add these ingredients, also add 1 cup of milk, or slightly more if the batter is too stiff after the eggs have been beaten. Bake in a tube cake pan for 1 1/2 hours at 325 degrees.

To freeze, wrap the cake in aluminum foil, and be sure that the hole in the center is also protected from freezer burn. The cake will keep for months. When you plan to serve it, remove it from the freezer and allow to thaw at room temperature for 2 or 3 hours. If you like, you can remove the cake and cut it in half, then thaw half and rewrap the other half and return it to the freezer.

Pumpkin or Sweet Potato Pie

Another fine dessert that freezes beautifully in a precooked fashion is pumpkin or sweet potato pie or custard.

Start with 1 1/2 cups of cooked pumpkin or sweet potatoes and add 1 cup of sugar, a dash of salt, 2 tablespoons of flour, a teaspoon of cinnamon, and 1/2 stick of margarine or butter. Mix these together well, and then add 3 slightly beaten eggs, 1 tablespoon of vanilla flavoring, and 1/2 cup of milk—slightly less if pumpkin is very watery. Blend all ingredients well and then pour into a 9-inch pastry shell, uncooked. Bake at 425 degrees for 10 minutes and then lower the heat to 325 degrees and bake approximately 40 minutes more.

Wait until the pie cools and then wrap it in aluminum foil or slip into a plastic freezing bag and freeze until ready to be served.

FREEZING COOKIES

One of the very best ways to keep a good supply of cookies on hand is to bake a double batch. Serve half the batch and freeze the other half. The same is true of many other desserts.

To bake a good freezer cookie, choose any recipe that does not contain such items as whipped topping, and bake the batch of cookies. Wait for them to cool, and then make a row of about 10 cookies. Add rows until you have a neat stack. Separate the rows only by a sheet of waxed paper. When all the cookies are stacked, wrap the entire group in aluminum foil or slip the package into a freezer bag and store in the freezer so that the cookies will not be broken by heavier packs of food.

Among the many cookies that freeze well are oatmeal, macaroons, chocolate chip, peanut butter, sugar, date-nut, and numerous other types. Some recipes, such as sugar cookies and chocolate chip cookies, can be made, and the dough can be frozen well. When you are ready to bake, allow the dough a few minutes to thaw, then slice the dough, place the slices on a cookie sheet, and bake as usual. It is a good idea to form the dough into a nice, neat roll about 1 1/2 inches in diameter and about a foot long, so that when you slice off cookies, they will be in the correct size to bake.

MAKING ICE CREAMS, ICE MILKS, AND SHERBETS

You can freeze two types of frozen desserts. You can make traditional freezer ice cream, turning the crank until the ice cream is of a proper consistency; then eat as much as you and your family want, putting the remainder in freezer cartons and store in the freezer.

You can also mix the ice cream ingredients and place the liquid in the freezer and allow it to freeze. If you choose the latter method, you may have to take the ice cream out and put it through the blender or you may need to beat it so that it retains an adequate supply of air. As you may realize, many commercially prepared ice creams have as much as 50 percent air in them. It is necessary to allow air room for expansion when the freezing occurs, and some companies even pump air into the freezing mixture to assure a better, more even consistency.

We recommend that you prepare a mixture of ice cream, freeze it in your home ice cream freezer, and store whatever isn't eaten immediately.

When you are freezing the ice cream, use ice that is in small fragments or crushed ice. Keep the freezer unit turning as you add ice, if you can. Every 4 inches of ice should have a generous sprinkling of salt poured on it.

When you start the rapid turning of the freezing unit, wrap a towel or other cloth around the freezer body to insulate against the heat of the day. Old newspapers, several thicknesses deep, also work very well. Before you actually start to make the ice cream you can wrap the freezer and use masking tape or Scotch tape to hold the newspapers in place. When the ice compartment is filled with ice and salt, spread a thick towel of cloth over the top to keep the cold air inside the freezer area and shorten the freezing time considerably. When the ice cream is ready to eat, the insulation will prevent thawing or melting while you serve the ice cream.

When you begin to serve, remove the dasher quickly and clean off the ice cream. After you have served the ice cream, replace the cover to the freezer tank and return the towel to the top. If you plan to freeze several cartons, you can dip out the ice cream to be frozen first, fill several plastic freezer cartons with it, and store in the freezer.

To be more logical, have the freezer cartons already in the refrigerator freezer unit so that the cold ice cream will not start to melt when it comes in contact with the freezer cartons.

If you need to watch calories, you can substitute regular milk for cream. You can also mix instant low-fat milk with the regular milk to cut even more calories. the exchange will cause the ice cream to become ice milk and it will have a slightly greater tendency to melt quickly and to freeze harder. It will also not be quite as fluffy as regular ice cream.

Ice Cream

You can make the ice cream mixture by your favorite recipes and you can add whatever fruits or nuts you like. For a good basic ice cream, you may want to try the following recipe.

Make the ice cream mixture the day before you want to freeze it, then leave the mixture in the refrigerator until you are ready to freeze. The time in the refrigerator creates a chill, resulting in a smoother, less airy ice cream. If you are using a hand-cranked freezer, use more salt than you would if you used an electric freezer. As suggested, crushed ice produces the best, fastest, and most evenly distributed freezing. If you can not obtain crushed ice, however, you can freeze large juice cans full of water. Place the cans in the freezer and leave until needed. To extract the ice you can hold the can by the bottom and tap the sides with a small hammer. The chips of ice will flake out quickly. You may have to break some of them in order to use them, but you have handy, free ice to use.

To make the ice cream mixture, start with 2 cups of milk and 4 cups of light or table cream. Scald the milk but do not boil it. Add the 1/4 teaspoon of salt and 1 cup of sugar to the milk and stir until the salt and sugar are dissolved. Next, add 4 tablespoons of vanilla, then cool the mix until time to freeze.

You may want to scald only the whole milk and save the cream to add after the sugar has been dissolved. You are now ready to put the mix into the freezer and make ice cream.

If you want to make a fruitier ice cream, you can crush 1 cup of strawberries, 3 or 4 bananas, peaches, or other fruits.

For an old-fashioned peach ice cream, blend 4 eggs, 2 cups of sugar, and 1/4 teaspoon of salt. Add 1 quart of scaled milk and cook in a double boiler until the mixture will coat a spoon. Remove from the heat and chill.

Next,, add 1 cup of whipping cream, 1 tablespoon of vanilla, 1/4 tablespoon of almond extract, and 2 to 3 cups of fresh, well-ripened peaches that have been chopped finely or crushed and sprinkled with sugar just before using.

Sherbet

Sherbets are summer favorites and can be kept for long periods of time. To make sherbet, use 1 cup of sugar and 3 cups of water, or less, if you use fruit that has its own juice. If you add a cup of juice, reduce the water content by 1 cup. Add 1 cup of crushed pineapple, cherries, or other fruits, 1/4 cup of lemon juice and pulp, and 2 egg whites that have been beaten well. After you boil the sugar and water together, and after the sugar is dissolved, add the other ingredients. Then place in the refrigerator to chill until you are ready to freeze it. You can freeze it in the ice cube trays of your refrigerator, then, after it is frozen, break it apart and beat with an electric beater until all the chunks and lumps have disappeared. Freeze it again and serve when well-frozen.

There are many other variations of the sherbet recipe, and you should use your own favorites. When the sherbet is frozen, you can leave it in the refrigerator and eat it soon afterwards or you can pack it into freezer cartons and store in the freezer.

12

Making and Freezing Breads

It has been said that bread is the staff of life, but the adage was much closer to the truth several years ago when home-baked breads were much more commonplace than they are now.

Today it is much easier—and, to be truthful, about as economical, if not more so—to buy bread at the grocery store. There are many delicatessen breads that are tasty, expertly prepared, and nutritional. The supermarket breads, too, are more often genuinely good.

If you wish to bake your own bread, however, you can do so without too much trouble. It does take time and effort, but it is well worth it to enjoy home-baked bread.

The breads you bake can be frozen. For loaves, wrap the entire loaf with aluminum foil and make sure that there is no way that moisture can be absorbed while the bread is in the freezer. If you prefer, you can use plastic freezer bags.

FREEZING BISCUITS

The breads you prepare can be frozen either as dough or as fully baked bread, and the bread can be as simple or as complex as you wish. For a really simple bread to freeze, try basic biscuits.

Basic Biscuits

To bake biscuits the easy way, start with 2 cups of self-rising flour and cut into it two egg-sized lumps of shortening. Add enough milk to make the dough moist and workable but not sticky. If you put in too much milk, you can always add extra flour; similarly, if the dough is too dry, add extra milk (Fig. 12-1).

Handle the dough as little as possible, and when it is properly mixed with the shortening and milk, dump it onto a well-floured surface, such as a biscuit board, counter top, or table surface. Sprinkle flour lightly over the dough and pat the flour into the surface lightly so that you can handle it easily. Slowly pat the dough into a flat oval no more than 1/4 inch thick. (Fig. 12-2).

Next, cut the biscuits by using a biscuit cutter or one end of an empty can. A soup can works well

Fig. 12-1. Keep the dough at a moist but not sticky consistency. Flour the board and the dough before kneading it.

Fig. 12-2. Roll the dough out until it is about 1/4 inch thick. Then sprinkle flour over the surface before you start to cut biscuits.

once the end is trimmed cleanly of all metal edges. Dip the cutter into flour so the flour won't stick to it and cut out biscuits 2 to 3 inches in diameter. Place these in a row and barely touching on a sheet of waxed paper. Fold the waxed paper over the row and add a second row. Continue the process until you have all the dough cut and packed (Fig. 12-3). Then wrap the entire pack in aluminum foil or slip the package into a freezer bag.

It may be handy if you make, as you did cookies, a double batch of dough each time you bake, and then bake half the biscuits and freeze the other half. If you do this for several days, you will soon have a nice store of biscuits in the freezer, where they will keep up to 8 months and perhaps longer.

When you need to bake the frozen biscuits, you can unwrap the package, pull the biscuits apart gently, and place them in a baking pan so that there is a distance of 1/8 inch between them. Bake them in a preheated oven set at 425 or 450 degrees for 8 to 10 minutes.

If you prefer, you can bake the biscuits as you prepare the dough. Then the baked biscuits can be frozen in the same manner that you froze the dough. When you want to serve, remove them from the freezer and place in the oven set at 300 degrees. Warm until crispy brown and thoroughly heated to the center. You may also wish to try sprinkling the frozen bread with water just before you heat. Usually, though, frozen bread does not tend to dry out; the opposite, in fact, is often true. Bread that is stale and dry can be frozen; when it is thawed there is seldom a trace of staleness.

For the reasons listed above, many people buy day-old bread and use it at home as soon as possible. Others freeze the bread and enjoy it as if it were fresh bread. Some people object to the idea of buying day-old bread; yet the same people often buy a fresh loaf, take it home, and don't use it for 2 or

Fig. 12-3. Use waxed paper to keep the biscuits separated, and fold the paper over each row. Wrap the entire batch of bread and freeze it in a freezer bag.

3 days. A fresh loaf sells for almost a dollar, while day-old bread is often on sale for 25¢, even less, per loaf.

Buttermilk Biscuits

If you want to bake buttermilk biscuits, follow the first recipe and substitute buttermilk for the milk. If you don't have buttermilk and don't want to go shopping, put 1 tablespoon of vinegar into a cup of whole milk and stir. It is ready to use instantly.

CORN BREAD

You cannot freeze corn bread until after it is baked, but baking is very easy and equally rewarding. Mix 3 cups of corn meal, 1 cup of flour, not self-rising, 1/2 teaspoon of soda, 2 teaspoons of baking powder, 1 tablespoon of sugar, 1 tablespoon of salt, 1 cup of shortening, 2 cups of sour milk (milk with a spoonful of vinegar), and 1 cup of water. Mix together well and pour into a greased pan. Bake at 350 degrees for 30 minutes.

An even simpler corn bread consists of 1 1/2 cups of corn meal, 1/2 teaspoon of soda, 1 teaspoon of salt, 1 tablespoon of melted shortening, and 1 cup of buttermilk. Mix well, pour into a greased pan, and bake at 400 degrees for 20 minutes or until crust is golden brown.

To freeze the corn bread, allow it to cool and then freeze the entire cake or slice it into individual servings and store the way you did biscuits.

The advantage to cutting the corn bread into servings is that it becomes very easy to take out only the exact amount desired for a meal. If you thaw the whole cake, you can still refreeze the portion that was not eaten.

Hot Corn Bread

Prepare corn bread by your favorite recipe, but before you put the bread into the oven chop up 1 cup of ham bits and 2 or 3 hot peppers. Mix into the batter; then bake as usual. The hot corn bread becomes almost a meal in itself.

Crackling Bread

For one of the most delectable Southern treats, try baking crackling corn bread. Cracklings can be bought at the meat market, but do not let the butcher persuade you that cracklings are the same as chitterlings, also known as chit'lins. Cracklings are the crispy brown parts left after lard has been removed by frying or rendering hog fat. While not noted for their lack of cholesterol, cracklings are wonderfully good in corn bread.

To bake this bread, you need 1 quart of corn meal, 1 pint of buttermilk, 1 teaspoon of soda, a pinch of salt, 1 pint of brown cracklings, and some warm water.

Crush the cracklings with a rolling pin or thick drinking glass, heat them, and stir the pieces into the dough, which should be thick enough to mold into smaller cakes or pones. If it is too thick, you can thin it with the warm water. Mold the dough into long, oval pones and bake at 450 degrees until brown.

You can eat the pones hot, or you can wait until they cool, wrap them in aluminum foil or place them in a freezer bag, and keep them frozen for 6 to 8 months. To use, simply remove them from the freezer and put them into a warm, 300-degree oven until they are heated to the center.

Grinding Your Own Meal

You can buy corn meal at the grocer's, of course, but you can also grind your own if you possess a small, portable grinder. They are available at many hardware stores and mail order houses for about $100. They are extremely useful for more than just corn meal; you can grind virtually any type of natural and hard food in them.

If you have good corn left in the field after the peak tenderness and ripeness has passed, leave the corn until it hardens on the cob or pull it, shuck it, and let it dry under controlled conditions. When the corn is dry enough so that you can shell it from the cob easily, you should shell and then dry the corn in your dehydrator (Fig. 12-4). You could instead spread it in the sun and cover the corn with a piece of cheesecloth or other insect-discouraging material until the kernels are thoroughly dry.

When the corn is dry enough, put a handful or two into the grinder and loosen the burr governor

Fig. 12-4. A homemade corn sheller can be made from a length of leather an inch wide and a length of slender wood. Make a slit in each end of the leather an inch from the end and run the point of the stick through the slit. To shell the ear, press the sheller against the rows of kernels and pull toward the palm of your hand.

so that the machine will grind the corn in a very coarse fashion. You can grind a hopperful in about 1 minute on this setting.

Reset the burrs to grind finer, and run the cracked corn through again. This time the grinding time for a full hopper is about 2 minutes. The third time set the burrs on the very fine position and grind the corn carefully and slowly so that you will get meal so fine you can rub it between your fingers and not feel any coarse elements. You can store the corn meal in the refrigerator or freezer, or you can place it in a plastic bag or metal container and make certain that it remains air-tight.

You can grind enough corn meal to last a typical family for a week in a matter of 10 to 15 minutes or even less, and the corn is not subjected to preservatives and other additives that supermarket meal contains.

Your mill can be used also for grinding wheat, rye, barley, soybeans, and virtually any and all types of additional grains.

To grind flour, secure a quantity of wheat, and, if it still contains the chaff, rub a handful between your palms until the other covering falls away and leaves only the grain seed. If the container of wheat is filled with chaff, hold the wheat grain over a container and let the wheat fall from your hands to the container, which should be 2 feet below your hands. If there is a good breeze blowing, the chaff will be blown away and the heavier grain will fall into the container. If there is no breeze, you can use a fan near the work area.

When the wheat is clean, you are ready to grind it. Like corn, it can be ground quickly if you run it through the mill on a rather coarse setting. The coarse setting for wheat must be much finer than the coarse setting for corn, naturally, since wheat is smaller. Grind it a second time on the very finest setting, and you will have a nice supply of wheat for your near-future baking.

WHOLE WHEAT BREAD

To bake whole wheat bread, you must have ample spare time and patience. First, empty 2 packets of baker's yeast into a pan of warm water—not hot but more than tepid. If the water feels only slightly cool to your flesh, then it is probably the right temperature; it should be about 10 degrees cooler than your normal temperature. While the yeast is reacting to the water, you can, in another pan, mix 2 tablespoons of granulated sugar, 2 teaspoons of salt, and 4 cups of bleached or white flour. Now mix the 2 cups of warm water and yeast into the flour and other ingredients. Knead the mixture until it is smooth and even.

With the dough in a warm place—on top of the refrigerator or on a range where the oven has been recently used—wait until bubbles start to form as the carbon dioxide escapes from the yeast mixture. It takes an hour or more for the bubbles to form fully.

Now take 1/2 cup of boiling water and mix it with 1/2 cup of brown sugar and 3 tablespoons of shortening. When the mixture has cooled, add it to the dough. Next add 4 cups of whole wheat flour and knead it into the dough.

Place the dough in a covered bowl and wait until it doubles its size. Take the dough, divide it into two equal lumps, and form the lumps into a ball and cover again. In about 15 minutes roll the dough in each ball into a flat oval, then roll the dough into a long roll or cylinder. Be sure to flatten the ends tightly so that there is an actual barrier formed of the pressed dough.

Bake the dough at 350 to 375 degrees for about an hour. If you prefer, you can freeze the dough rather than bake the finished bread. For a terrific modification, rather than use sugar use 1/2 cup of strained honey. You can also use 1/2 cup of condensed milk rather than all water.

To freeze the bread, wrap it as you would wrap any other bread and store it in the freezer. You could also put it into freezer bags and seal tightly.

MUFFINS

You can bake muffins using 1 egg, 1 cup of sweet milk, 2 cups of flour, 4 teaspoons of baking powder, 1/4 teaspoon of salt, 2 tablespoons of sugar, and 2 tablespoons of melted shortening.

Beat the egg, add the milk, then the flour and

the rest of the ingredients. Mix well and put into a greased muffin pan. Bake about 20 minutes in a moderate oven.

You can modify the recipe for basic muffins in many ways. One way is to pour 1 cup of sour milk over 1 cup of oatmeal and let the mixture stand while you prepare other ingredients. Add 1 unbeaten egg to the oatmeal and 1/2 cup of brown sugar and mix into preparation.

Last, add 1 cup of flour, 1 tablespoon of salt, 1/2 teaspoon of baking powder, and 1/4 cup of melted shortening. Mix all of the ingredients well and pour into a greased muffin pan. Bake for 20 minutes at 400 degrees.

Instead of oatmeal you can add cranberries, nuts, and a wide variety of similar foods.

You can store the baked muffins easily by placing a row of muffins on a sheet of waxed paper; then bend the paper over to cover the tops of the muffins and add another row. Pack the entire group into a freezer bag or wrap in aluminum foil.

ROLLS

Start baking rolls by emptying a package of yeast into 1 cup of water about 85 degrees. While the yeast reactivates, mix 2 cups of plain enriched flour, 1 cup of whole wheat flour, 1 teaspoon of salt, 2 tablespoons of sugar, and 3 tablespoons of corn syrup. Add 3 large tablespoons of shortening, mix well, and add the yeast and enough water to make a soft dough. After you knead the dough well, set it aside to rise. Keep the exterior of the dough greased lightly or it will start to crack and dry.

Wait for 2 hours until the dough has doubled its size. Then put it in loaf pans and allow the dough, which you should knead again, to double in volume a second time.

Shape the dough into loaves and bake in a preheated hot oven for 15 minutes up to 45 minutes, depending upon the thickness of the rolls or loaves.

You can freeze the dough after it has risen the second time, but be sure to grease the outside again lightly before you wrap the loaves in aluminum foil. If you bake the bread, you can freeze it as you would any other bread.

NUT BREADS

For a very basic nut bread, combine 2 well-beaten eggs, 1 cup of sugar, 1 cup of sweet milk, 3 cups of enriched flour, 1 cup of walnut or pecan meats, and 1 teaspoon of baking powder. Bake the loaf in a moderate oven for about 35 minutes or until well done.

A Variation

For a variation of the above bread, mix 2 tablespoons of butter, 1 cup of sugar, and 1 cup of chopped dates. Now pour a cup of boiling water over the above mixture and allow it to stand until cool. Add 1 well-beaten egg, 2 cups of flour, 1 teaspoon of soda, and 1/2 cup of chopped nuts.

Mix the ingredients thoroughly and shape a loaf from the dough. Bake for an hour or slightly longer at 350 degrees.

Other Breads

Banana bread can be made by following much the same plan of action. Use 1/4 cup of butter, 1 teaspoon of lemon, 1 teaspoon of vanilla, 2 teaspoons of soda, 1 cup of finely chopped nuts, 2 cups of sugar, 4 cups of flour, 2 cups or slightly less of crushed bananas, and 4 eggs. Mix the ingredients well and bake at 350 degrees for about 1 hour.

In a similar fashion, you can follow your own recipes or modify existing ones. You can add raisins, plums, orange pulp, apples, or other fruit or berries—blueberries are great for this kind of bread, as are currants—and stick to the basic mixture of flour, sugar, eggs, and spices.

Freeze these loaves either unbaked or fully baked. They will keep, if properly wrapped, for well over 6 months in the freezer. To use the frozen loaves, remove from the freezer, allow to thaw for an hour, then heat in a moderate oven. When you slice them use a knife with a serrated blade or a warm, greased knife. If you froze unbaked loaves, do not thaw; simply bake at 400 degrees for approximately 1 hour.

13

Preserving Meats, Poultry, and Fish

Of all the foods that can be preserved, few are more challenging than meats, poultry, and fish. One simple and easy way is by freezing, but the other methods require considerable time and effort. In some instances, notably for curing, the final product is better than fresh meat; in freezing, the taste remains nearly the same; in drying, there is a distinctively different, and often delicious, taste; in canned meats there is perhaps a dimunition of flavor and texture, but this difference is likely attributable to the quality of the meats or poultry selected for canning.

Several other considerations should be stated early. There is a distinct danger of food spoilage if you do not properly care for the meat. The necessary attention must be toward working room temperatures, storage temperatures, specific preparations of the meat, maintenance of the storage temperature, and fairly exact measurements for curing and drying procedures and ingredients.

With fruits, vegetables, and berries, for instance, there is only a blanching period, not a curing period at all, but with many kinds of meats—beef, venison, veal, and others—there is a virtual requirement of a five-day curing period during which time the room temperature must be held below 40 degrees.

Again, with fruits and vegetables, if a jar spoils, you can replace the contents of the jar by picking a few more beans or, if you must, by paying a few cents at the store. If you have an entire ham spoil, you cannot usually obtain another ham quickly and easily. You are also talking about quite a large amount of money and dealing with long, hard hours of work with the ham, starting with the feeding of the hog, and continuing through the slaughtering, dressing, and all of the other work that goes into the production of a high-quality ham.

On the opposite side though, consider the astonishingly great taste of properly cured ham as well as all of the other fine cuts of meats that can be obtained from the hog: bacon, sausage, livermush, tenderloin, pork chops, roasts, and numerous other lesser values in nutrition and taste.

In terms of the importance of the animal to your diet, food budget, and gustatory delights, therefore, exert careful attention to all matters concerning the preparation of the meat, poultry, or fish for home preservation.

PREPARATION OF FOOD

If you buy your meat already slaughtered but not cured, you must be prepared to take care of the meat in the most reasonable ways. If the source of the meat is a fairly long drive from your home—an hour or more—you should make provisions for keeping the meat properly cool during the trip home. These provisions include taking along a quantity of ice or various forms of insulation, or, best of all, choosing a day that is very cold so that the meat will not be exposed to temperatures that are too high. Meat should never, unless totally unavoidable, be exposed to prolonged temperatures over 38 degrees.

Once you get the meat home, chill it immediately. Once the meat has been properly chilled, take care to keep the temperatures low while you work with the meat.

When working with pork, give your earliest attention to those portions of the hog that are either most perishable or most valuable to you. You will want to handle the hams, shoulders, chops, loins, and roasts very quickly, and then concentrate on the sausage and livermush, with pickled pigs' feet probably coming at the end of the processing period. Many people really don't care whether they get their souse meat and pigs' feet processed or not.

The beef is similar. Steaks come first, for most people, with roasts and hamburger following.

Whatever the type of meat, get rid of all the parts you do not intend to save as quickly as you can. Cut off fat, gristle, tendons, and other undesirable elements and discard. In the meat preparation, always remove all vestiges of viscera, and trim away any spot that has been bruised or that has congealed blood in it.

To prepare wild small game, first remove all fur and skin to which the fur is attached, and slit the animal down the stomach from between the front legs to the back legs. Pull out and discard all intestines. Next, cut off the head and all four feet. Then, after washing the animal carefully, use a sharp knife and follow the outline of the legs and cut off the entire leg, including the very thick area where the leg joins the body. Do this with all 4 legs. Then cut the back into 2 right behind the rib cage. Your best bet is to process the legs and plan to cook and eat the lesser sections, since the rib cage and lower back portions take up a great deal of canning room and actually provide very little meat.

If you buy your meat from a wholesaler, you can have him prepare the meat for you. He will not only do all of the slaughter work but will cut and package all of your cuts.

You can also buy just the beef half or quarter, take it home, and do the carving and trimming yourself to save some money. Usually the minimum price for the butcher's work is 10¢ per pound, which doesn't sound like very much until you consider that there may be 200-300 pounds of meat. Often the price is much higher, even up to 25¢ and more per pound.

Some foods, such as wild animals, wild fowl, fish, and even some domesticated animals have a "wild" taste. To eliminate this undesirable taste, cut the meat into portions small enough to fit into a deep dish or bowl and soak the meat in a brine solution for at least 8 hours.

If you have a pair of rabbits, for instance, cut off the legs and trim the rabbit carcasses as you wish to have the pieces cut for cooking, and place the meat pieces in a bowl. Now sprinkle or pour a generous amount of salt over the flesh—about 1/4 cup or less for two rabbits—and then add mildly warm water. If you prefer you can dissolve the salt in water before you pour the water over the meat.

After the meat has soaked for an hour, pour off the water, rinse the meat, and add a second brine solution of 4 tablespoons of salt for a quart of warm water and let the meat soak the remainder of the 8 hours. Then pour off the salt water, rinse the meat well, and drain it for several minutes. Blot it dry with paper towels or cloths, and prepare to precook it for canning or wrap it in foil for freezing.

CANNING

Meats of all types as well as fish, chicken, turkey,

and assorted seafoods are low-acid or neutral, thus they must be canned only by pressure canning methods. Do not attempt to process any form of meats in water-bath canners.

It is obvious to anyone who has been interested in home food preservation for any length of time that thousands of rural people have canned meats in a water bath for decades, but the danger of botulism and other food poisoning is too great to run such an unnecessary risk. The rural folk of generations ago had no pressure canners, so they had to use water-bath methods if they canned meats at all, and they learned to be experts in the handling of the flesh prior to processing, as well as during preparation to eat the meat.

Preparation

All canning can hope to do is preserve the food so that it is tasty, nutritional, and safe. For the most part, canned foods taste almost as good as the fresh and at times at least as good as fresh foods, but in order to have good results with canned meats and poultry you should start with high-quality flesh.

If you raise your own meat, you can see to it that your animals are fed only good foods. You can also be certain that the animals have had no health problems. If you buy your meat, you must depend upon the person's reliability and truthfulness concerning the care of the live animal. It would be very wise, if expedient, to visit the farm or slaughterhouse to make at least a cursory visual inspection of the general premises. If you buy government inspected meats, you are reasonably sure of getting high-quality foods.

For canning purposes, cut the meat off the bone also, since bones take up considerable space in a canning jar and add little if anything to the final product.

Put the bones and attached small scraps of meat into a large pot and start them to boil so that you will have abundant broth. To process broth add at least 1 cup of water for every cup of broth and process in a pressure canner for 1 hour at 10 pounds of pressure.

The meat that is being prepared can be cut to give you adequate servings for your family and to save space in the jar, which should be packed fully and tightly with at least an inch of space left at the top of the jar. You will not need any liquid in the jars.

You may pack meat according to the raw-pack method or by the hot-pack method. If you plan to use raw packing, no further preparation is necessary after the meat has been properly cut and cleaned.

For hot-pack processing you should precook the meat long enough for the heat to reach the center of the cuts. The thicker the cuts, the longer the cooking time will be. If the meat is not particularly tender, you can cook it in a pot partially filled with liquids. You do not need to boil the meat but the temperature should be high enough so that the water is near the boiling point.

For tender cuts of meat you may use the oven, set at 200 degrees or the lowest setting, for precooking. Leave the meat in the oven for 1/2 hour or slightly longer, then prepare the jars for packing. All jars should be hot and sterilized.

Processing Times

It is best to can meat in either quart or pint jars, and in some cases the smaller amounts are better. Fish, for instance, are sometimes hard to can handily and should be canned in pints.

For beef, mutton, and pork, process all types of cuts at 10 pounds of pressure for 1 1/4 hours for pints and 1 1/2 hours for quarts.

Fish should be processed at 10 pounds of pressure for 2 hours or slightly less. Take care never to let the pressure drop below 10 pounds.

Wild game of all types, including quail, pheasant, deer, elk, rabbits, squirrels, and wild hogs, should be processed in pints for 1 1/4 hours and in quarts for 1 1/2 hours at 10 pounds of pressure.

Hamburger and sausage should be hot packed and processed at 10 pounds of pressure for 1 1/4 hours for pints and 1 1/2 hours for quarts.

Sausage

To make sausage, you can grind in a good chopper all of the hog, or you can use selected portions, such as the shoulders or picnic hams. Most people use meat trimmed from hams, sides, loins, head, and anywhere else that lean meat or fat-and-lean mixed meat can be found.

Grind the meat with the chopper set on the coarse blade setting and feed the meat through. It takes quite a long time to grind all of the sausage in a hog, and you might want to consider taking the meat to your local butcher and paying him to run it through his grinder.

Still, if you persevere, you can get the job done. And when you have completed the grinding, you can add seasoning. You can buy special sausage seasoning for either mild, hot, or very hot sausage, or you can season to suit yourself.

A typical mixture is 6 ounces of salt, 1 ounce of black pepper, which is quite a lot of pepper when you consider how little ground pepper weighs, 1 ounce of red pepper, and 1 ounce of sage for each 25 pounds of ground meat. You can spread the ground meat on a cutting block or table and sprinkle the seasoning on the meat, then knead the meat as if it were flour to get the seasoning to spread uniformly through the sausage.

If you like, you can run the sausage through the grinder a second time to help spread the seasoning better. And if you like really hot sausage, you can increase the red pepper.

After you have seasoned the sausage, precook and pack into hot, sterilized jars. Use a clean, damp cloth to wipe off the rim of the jars and put on lids that have been placed in boiling water. When the lid is seated properly, add the rings, which also have been in the hot water. Tighten firmly by hand and put the jars into the pressure canner.

When making hamburger, grind the meat as you did the sausage and you can season it lightly with salt and pepper before packing and canning it.

Poultry

Poultry can be canned bone-in or boneless, and it can be raw-packed or hot-packed. After you kill or obtain the bird, cut off the head and hang the bird upside down to bleed. Let it bleed for several minutes, then dip it into scalding water to loosen the feathers. Pluck the feathers off as quickly as you can and then singe the bird's carcass to eliminate pin feathers.

You can singe with any type of flame that has little if any smoke. An old-fashioned way is to hold the bird over a blazing sheet of newspaper made into a funnel with the large end up. A more modern way is to use several spoonfuls of ordinary alcohol in a small container that you don't mind discarding after use. Set the alcohol on fire and singe the bird well.

Afterward slit the intestinal cavity open and remove all intestines and liver, heart, and other giblet foods. Cut off the oil sac that is found just above the tail and discard it. Wash the bird several times to be sure that all blood is cleaned away and then let the carcass drain for a few minutes before you soak it in salted water. Cut the bird apart by removing the legs first, then separate the legs into drumsticks and thighs. Cut off the wings and snip off and discard the wing tips. Cut the breast out whole, if possible, and also cut up the back sections and rib cage.

You may want to keep the bony parts of the poultry to cook separately and perhaps make chicken and dumplings.

Precook if desired or cold pack and pack into a hot, sterilized jar, close, and process at 10 pounds of pressure for 1 1/4 hours for pints and 1 1/2 hours for quarts.

Fish

Use only fresh fish and check to see that no offensive smell is developing. Fish can spoil very quickly, and you must see that the fish is never exposed to excessive heat for a prolonged period of time—1/2 hour at summer temperatures is dangerously close to being excessive.

Clean the fish well. Scale by using the edge of a knife, or you can use a spoon. Scrape the spoon, with the top facing in the opposite direction from the natural position of scales, and rake the spoon across the scales from the tail toward the head. The scales will be dislodged easily. After scaling, wash the fish quickly and drop small fish into a container of cold water that has about 4 tablespoons of salt dissolved in it.

After the scaling has been completed you can also cut off the head of the fish and disembowel it.

Wash the cavity carefully and make sure the fish soaks for 1/2 hour in the cold salt water.

Raw pack the fish in pint jars and close the lids. Be sure the lids and rings have been soaking in very hot water and that the jars themselves are sterilized and hot. Tighten the rings firmly and process the jars of fish at 10 pounds of pressure for 1 hour and 40 minutes.

FREEZING

If you are going to freeze your meat, there is very little preparation necessary. For beef, venison, mutton, veal, and similar meats there is a recommended 5-day aging period in which the meat is kept hanging in a cooler room, with temperatures never above 38 degrees. Aging is supposed to make the meat taste much better. We recommend letting the wholesaler age the meat, since it takes a lot of space to hang a beef for 5 days.

Beef and Pork

For beef, once the meat has been cut into desired servings, it is a simple matter to wrap the cuts in aluminum foil or put it into a freezer bag and slip it into the freezer.

When you first obtain the meat, chill it so that it is kept at a temperature of 34 to 38 degrees. Do not freeze and do not allow the beef to get higher than 38 degrees. If you have a large quantity of beef, keep it chilling for about 5 days before freezing it. If you buy the meat at the supermarket, it will have already been aged and it can be frozen as soon as you get it home. Veal can be aged for 1 or 2 days.

You can also freeze pork, mutton, and other meats as soon as you reach home. Only those meats and poultry and fish that are obtained from private sources should be pretreated before freezing.

If you want to precook your beef and pork, do so and then freeze. Large amounts of pork, unlike beef, should be chilled and frozen as quickly as possible; do not attempt to age pork except cured portions. Wrap your beef and pork in strong paper or plastic so that the package will not become torn or damaged in ordinary handling. The exposed meat, if the package is ripped, can be affected by freezer burn.

Another safeguard is to remove bones before packaging, when practical, so that jagged bone edges will not damage packing paper in storage.

Poultry

Poultry can be frozen equally well whole, halved, or cut up into individual portions. The advantages to freezing smaller portions are that you can take out as little as you like. If the entire turkey, on the other hand, is frozen whole, you have no choice but to thaw and cook the entire bird. You can easily freeze the leftovers after the turkey has been cooked and served.

To freeze poultry, prepare the bird according to the directions given in the canning section of this chapter. If you want to freeze the whole bird, make sure that it is washed clean inside and out and wrap it securely so that no portion is exposed to the air. If you want to freeze the liver and other giblets, wrap these separately, stuff them into the body cavity, and freeze the whole assembly together.

If you want to freeze half a bird, slit the abdominal cavity from the neck to the tail and lay the bird on its back and spread the cavity wide open. You will be able to see the spine or backbone of the bird. Connected with the spine are the rib bones. These bones can be cut easily just to the left or the right of the backbone if you lay the edge of a thick-bladed knife against the rib and push down firmly. You can feel the blade force its way through the rib. Cut the meat between ribs and move to the next bone. In this fashion you can cut or break all of the ribs and then use the blade to sever the thin layers of flesh that connect the two halves.

The whole operation should require no more than a minute, and you will have two portions of the bird to freeze. You can freeze the halves in one package or you can make two separate packages. If you freeze them together, wrap each half in waxed paper before wrapping the 2 halves in one package.

If you wish to freeze the chicken in parts, use the thick blade of the knife and force the cutting edge between the joints of the bird and exert gentle but firm pressure away from your body. Never cut by pulling the knife blade toward you; if the bone and meat should yield suddenly, the blade could do serious damage to your chest or arm.

You can select enough parts for your family needs for one meal and freeze these parts together, with each piece wrapped in waxed paper. If there are three members of the family and each person will need only one piece of chicken, then freeze, for instance, one thigh, one breast portion, and one drumstick in a package. Freeze all of the bony parts together and use them for chicken and dumplings or similar meals.

Fish

Freezing fish is very easy. If the fish is small, you can slit the abdomen open and remove the entrails and gills and then cut off the fins, head, and tail. You can freeze several small fish in one package.

Clean the fish as the instructions indicated in the first part of this chapter. After you have scaled the fish and cleaned out the entrails, remove the fins. Start with the large fin on the top of the back. Hold the fish with your fingers spread over the 2 fins just behind the head. If you need a firmer grip, use a paper towel to hold the fish steadily. With your other hand hold a thick-bladed knife at the base of the fin and on the same side as the tail. Curve the blade slightly so that it penetrates the skin just behind the fin. You can straighten the blade's direction, slip it easily under the fin, and trim off the entire fin in one easy movement.

Do the same with the fin on the fish's underside. You will now have the two fins near the gills, and you can cut these off easily by grasping one fin with your thumb and forefinger. With the fish lying on a cloth or paper towel, lift the fin so that you can see where it is joined to the body. Slip the knife blade under the skin slightly and cut the flesh surrounding the fin. The fin should then be free of the body. Do the same thing on the other side.

Holding the fish firmly against the work surface, place the knife blade just behind the fish's eye—between the eye and the point where the first fins were attached—and push down and start a gentle sawing motion until you feel the blade encounter the backbone. Push down harder until you feel the blade pass through the backbone, and then resume your gentle sawing motion.

If the fish is of a medium size—too large for a frying pan—run the point of a knife through the fish, starting at the base of the rib cavity, then push the knife upward until the point and part of the blade emerge on the upper side. With the cutting edge toward the tail, turn the knife flat and let the blade rest against the backbone and gently work the edge through the flesh toward the tail. Turn the knife and start cutting again just below the rib cavity, but this time the cutting edge is toward the head. Tilt the blade upward so the edge will glide across the rib bones and emerge just behind the head.

You can lift the filet off and place it to one side. Turn the fish over and repeat the process. When you have enough filets to make a good meal for your family, wrap the individual pieces in waxed paper and then wrap the entire package in heavy freezer paper or foil or place it in a freezer bag and store in the freezer.

If the fish is very large and you wish to have fish steaks, cut off the fish's head, slit open the intestinal cavity, and remove entrails. Start, after scaling the fish, behind the first fins and, using a sharp knife, cut straight through the fish. If the backbone is large enough to prove difficult to cut, you can feel the joints with the edge of the knife and cut through the backbone at the joints.

Each steak will be an oval about 1/2 inch thick. Lay one steak on a piece of waxed paper, fold the paper over to cover the top of the steak, and lay another steak on the paper. Continue the process until you have an adequate amount of steaks for a meal; then wrap the stack in heavy freezer paper or store in foil or freezer bags.

Eggs

You can freeze eggs, too, quite easily. Start by placing the eggs in a bowl of water. If you gather your own eggs, you will know how fresh they are; if you buy them, you can separate fresh from old eggs by the floating test. Crack one egg at a time and drop it into a saucer. If the egg looks satisfactory, put it into a larger container. Break each egg first into the saucer. If you should encounter a bad egg, discard the egg, wash out the saucer, and break the next egg. The reason for using the saucer, obviously, is so that, if there is a bad one, the badness will be confined to the saucer.

When you have broken all the eggs you plan to freeze, use a wooden spoon and beat the eggs slowly until all yolks are broken and mixed well with the whites. Then, using a spare refrigerator tray, measure out enough egg mixture to fill a tray compartment. Proceed to fill each compartment with egg mixture. When you have finished sprinkle salt and sugar over the tray, or you can use 1 tablespoon of sugar and/or 1/2 teaspoon of salt per cup of egg mixture.

When you are ready, freeze the entire tray of eggs. When the eggs are frozen hard, remove the tray from the refrigerator, turn the tray upside down, and run lukewarm water over the bottom until the eggs slip into a container held under the tray. What you have, in effect, is square eggs which can be frozen together in whatever number best suits your family's needs. If you bake several cakes at a time, you could pack enough eggs into one package to bake three cakes. If your family needs 6 eggs per meal, pack and freeze the group as one package. As you did with meats, clearly mark eggs on the outside of the package to facilitate finding them when you need them.

FREEZING MEAT, POULTRY, AND FISH DISHES

At times it is as convenient to cook the meat and freeze the dish to be used later as a main course or as a supplement to a meal. Among the main dishes that can be frozen are casseroles, meat loafs, pot pies, chicken and dumplings, livermush, and barbecue.

Livermush

To make livermush, cook 1 hog's liver and 1/2 of a hog's head until well done. When the meat is done, take it from the broth and cool it. Remove the skin, if any, and run the meat through a food chopper. Return the ground meat to the cooking broth, season with salt, pepper, sage, and other spices to taste, and thicken with corn meal. Stir constantly while the meat and corn meal are cooking. Cook slowly until the mixture thickens. You will need about 2 quarts of corn meal for a large liver and half of a head. Wet the corn meal before mixing it with the liver so the meal does not have a tendency to lump when it comes in contact with the hot broth.

After the livermush has thickened properly, pour it into a pan and allow it to cool. When it has cooled well, you can package the livermush in small cubes and wrap and freeze.

Fish

For a good baked fish, leave a fish whole, except for head and entrails. Make a dressing of bread crumbs, 1 egg, salt and pepper to taste, and dots of butter. Stuff the dressing into the fish and tie the fish cavity closed with twine. Bake for about 30 minutes at 350 degrees. Wait until the fish cools and then wrap and freeze.

A salmon loaf can be made from 1 pound of salmon, or any fish that is flaky and light, 1 cup of rolled cracker crumbs, 1 tablespoon of grated onion, 1 tablespoon of lemon juice, 2 tablespoons of melted butter, 1/2 teaspoon of salt, 1 beaten egg, 2/3 cup of milk, and a dash of pepper. Mix the fish and other ingredients gently. Flake the fish so that the portions will mix well with the other ingredients, and pack it into a well-greased loaf pan. Bake in a moderate oven for about 45 minutes. Let the loaf cool, then wrap properly and store in the freezer until ready to use.

Other Dishes

To make chicken á la king, start with 1 quart of

chopped chicken meat. Let 3 tablespoons of shortening melt in a pan and add 1 small chopped onion, 1 tablespoon of chopped sweet red peppers, 1 tablespoon of sweet red peppers, and 6 mushrooms (optional). Add 3 tablespoons of flour. Blend or stir together until well-mixed. Add 3 cups of chicken stock slowly and stir until smoothly blended. Cook until the mixture thickens, then add 1 cup of diced carrots and the chopped chicken. Cook until well done, remove from heat, and allow to cool. Then pour the mixture into freezer cartons and store in the freezer until needed.

Pot pies can be made from chicken, beef, pork, turkey, and other meats. Roll biscuit dough (see Chapter 7, on breads) to 1/2 inch thickness. Cook 2 cups of chopped meat until well done. Use the stock to make a thin gravy, and then add carrots, potatoes, and any other vegetables that you prefer. Season to taste with salt, pepper, and other spices. Spread dough, rolled thin, into pie pans, then pour the mixture of meat and vegetables into the pie pan. Add a top crust and then bake the pie in a moderate oven for 30 minutes or until done. Remove the pie from the oven and allow to cool. Then wrap it properly and store in the freezer until needed.

To prepare a basic meat loaf, mix 1 1/2 pounds of ground beef, 1 cup of tomato juice, 1/2 green pepper chopped finely, 1 small chopped onion, 1 teaspoon of salt, 1/2 teaspoon of black pepper, 1 teaspoon of baking powder, and 2 tablespoons of flour. Mix all the ingredients well. Shape into a loaf and bake in a greased pan until the outside is well browned. Lower the temperature to 350 degrees, add 1/2 cup of water, and cook for another 30 minutes. Remove the loaf from the oven and let it cool. Then wrap and freeze.

To make a barbecue dish that freezes well, cook 4 medium sized onions and 1 bell pepper in melted shortening. When the onions and pepper are done, add 2 pounds of ground beef or ground pork. Add a teaspoon of mustard, 2 cups of tomato ketchup, 1 tablespoon of sugar, seasoning to taste. Mix well and cook until the meat is done. Then remove from heat, cool, wrap well, and store in the freezer.

For a good spaghetti dish to freeze, use the recipe given in Chapter 9. Cook a package of spaghetti in salted water and pour the sauce over the cooked spaghetti. When cool, the dish can be poured into freezer cartons, sealed, and frozen.

There are many other similar dishes that can be frozen. When you are ready to use one of the dishes, remove it from the freezer and prepare to serve it as directed below. If you freeze the dishes in aluminum pans, you can also heat or bake them in the same container. If you froze the dish in a plastic carton, you can allow the food to thaw at room temperature until you can pour it from the carton. You can also run hot water over the bottom of the carton or let the carton sit in a bowl of hot water until the food can be removed easily.

COOKING WITH FROZEN MEATS AND OTHER DISHES

Using the frozen meats is simple. When you want a steak, simply remove the package and open the wrapping so that air can reach the meat. Let it thaw for an hour or so, then broil, fry, or bake as you wish. If you know in the morning that you will be using the dish that night for dinner, remove it from the freezer and store it in the refrigerator until evening, at which time it should be thawed sufficiently to cook.

Such a dish as baked fish or meat loaf can be put into the oven while the dish is frozen. The dish will thaw and heat well and evenly if you use an oven setting of about 300 degrees.

Barbecue can be heated in a saucepan or double boiler, and livermush can be sliced while still frozen and fried until it is crispy brown, then served.

The pot pies can be put in the oven while they are still frozen and baked or heated.

Any frozen meat or chicken can be thawed in a pan of lukewarm water, left on the counter to thaw, or cooked while still frozen.

If you like, you can freeze the whole meal and then cook or warm it all at once. You can, for instance, take an aluminum pan with various food compartments in them and freeze a generous serving of steak or roast, a serving of green peas, some mashed potatoes, and even a roll simply by wrapping all of the pan or plate in foil and storing it in your freezer.

When you want to eat it, place the whole pan, foil and all, in the oven and set the temperature on 300 degrees. Warm for 20 minutes or slightly less—long enough for all of the food to warm throughout. This type of whole meal, which can have literally dozens of variations from fish to barbecue to an all-vegetable meal, can be frozen cooked or uncooked, but cooked is better since you don't have to worry about cooking the various foods for different times and under different conditions.

Obviously you couldn't use mashed potatoes, but you could use several other vegetables that could be baked while the meat is cooking. But, as we recommended, previously cooked foods are excellent for this type of freezing. A good idea is to cook an entire roast, for instance, and the vegetables to eat with it. Carve the rest of the roast, after the initial meal, and portion out the rest of the vegetables as individual dinners to be frozen and eaten later. Even if you can freeze only one roast dinner, later you might add a chicken dinner, a fish dinner, or other kinds of frozen entrees, and then your family can have its choice.

DEHYDRATING MEATS

Dried beef, also called jerky, can be prepared in a simple fashion. First, since fat will not dry well, trim off all fat, then slice the meat into thin strips that are about 1/4 inch thick or less. Do not leave any bone, tendon, ligament, or other tissues in the meat. Do not attempt to dehydrate pork, since pork flesh is totally unsuitable and may be harmful to your health.

Once you have sliced your meat thinly, you can soak it for 24 hours in a salt water brine made by dissolving 1 pound of salt in 1 gallon of water. You can add as many other seasoning ingredients as you like, such as brown sugar, pepper, and garlic.

You can brine or spice the meat in a simple mixture of the salt, sugar, and spices that you prefer and literally coat the meat with the seasoning. Put a thin layer of seasoning in the bottom of a container and spread the meat over it, then cover the meat with a second thin layer of seasoning. Proceed to make as many layers as you have meat and seasoning to supply.

When you remove the meat from either the brine or the dry seasoning you can dry it off or rinse it, depending upon whether you want to have heavily seasoned jerky or mildly seasoned meat.

You may precook the meat before drying it, if you like, and make your job much easier. Leftovers, for example, can be chopped finely and placed in a dehydrator tray and dried until ready. You can tell when meat is dehydrated enough since it will actually break in half if bent completely double.

It is not a good idea to brine or dry-season in salt alone since salt can tend to make red meat stringier and tougher.

If chicken, turkey, and other non-red meats or types of flesh are already cooked, they may be dehydrated and stored as you would any other dried foods.

One important note: when slicing meat for jerky you want the jerky to dehydrate in long strips; so cut it along the grain.

If you have dehydrated chicken or turkey, you can also chop it extremely finely after dehydrating and later use it in soups or stocks. The dehydrated meat will add excellent flavor to soups. You can even grind it in your good chopper if you prefer, because you do not want large pieces of meat for the soups. Remember that you will reduce the weight and volume of the meat by 80 percent. If the chunks of meat are left too large, they will dry very poorly and when added to water the meat will reconstitute and swell until it is larger than a desirable size for soup.

If you sun-dry jerky, it will take 3 or 4 days for the meat to be properly dried, and you will need hot, cloudless days. If it rains, you will have to take the meat indoors until the weather clears.

In the oven you can dehydrate meats as you did the various vegetables. Spread the meat on the oven racks and set the oven on "warm" or the lowest setting. You want to dehydrate, not cook, the meat. Leave the oven door open 3 inches for the entire process.

If you have a small fan that can be set up near the cracked door, turn the fan on and run it throughout the dehydrating, or at least for 15 minutes or so every hour so the moisture-laden air can escape easily.

Since the meat juices will drip during the dehydrating process, you may want to use aluminum foil in the bottom of the oven.

If you wish to dehydrate fish, you can do so by cleaning the fish well, washing thoroughly, and then soaking the fish in salt water for 12 hours or slightly longer. Remove the fish from the brine, blot dry, and spread it on the dehydrator racks. We recommend drying fish in the oven or in the dehydrator, since fish has a tendency to spoil very quickly in the hot sunlight. The salt, of course, retards spoilage, but the faster the drying is accomplished, the better.

With all dried meats, you can add as much sweetening as you wish by putting brown sugar into the seasoning, and you can sprinkle brown sugar on the dehydrating meat.

14

Making Cheese and Yogurt

When milk sours, curd is formed, and a watery substance remains. The watery substance is whey. In its simplest sense, cheese is made by coagulating milk by using rennin, acid, or both and then by removing the whey and preserving the curd.

The first step in making cheese is coagulating the milk. The rennin used is extracted from the fourth stomach of a milk-fed calf. Rennin causes a change in casein, which is a phosphoprotein that is one of the chief constituents in milk. The casein coalesces and curds are formed. Later the curd is cut to release the whey, or at least a part of it; some of the whey will always be left in the curd.

When the curd is formed and removed from the whey, you have the first stages of cheese, which is referred to as fresh or green cheese. As it grows older cheese may be said to ripen, age, or cure.

As the cheese passes through the various ripening stages, many changes occur, depending on climate, the amount of moisture content, and acidity. By the time the finished product is obtained, 9/10 of the volume of the milk has disappeared. The final 1/10 is the cheese. So from 100 pounds of milk you can get 10 pounds of cheese. Since the watery part of the milk was extracted and only the richer, heavier parts remain, you can see why cheese is so calorie-laden and nutritious.

Among the more frequently eaten types of cheese are blue, brick, camembert, cheddar, cottage, cream, edam, gouda, mozzarella, parmesan, and processed cheeses.

Blue cheese is a mold-ripened cheese made from cows' milk. The cheese is heavily salted and it is aged from 3 to 6 months usually.

Brick cheese is semihard, slices well without crumbling, and is ready in several weeks.

Camembert cheese, also from cows' milk, is mold-ripened and has a harder outside crust and a soft inside texture.

Cheddar cheese is a smooth-textured cheese with only a few holes and is among the most popular types of cheese.

Cottage cheese is soft and unripened cheese that will keep only for a few days under refrigeration. Cream cheese, too, is unripened and will keep under refrigeration only a few days.

Edam cheese is ripened until it is hard and smooth. Gouda is semihard or hard cheese with a flavor much like that of edam cheese. Mozzarella cheese is soft and unripened. Parmesan cheese is soft and is ripened for 1 to 2 years. Processed cheese is usually a mixture of several types of cheeses and will generally have considerably less flavor than the pure cheeses.

MAKING COTTAGE CHEESE

To make cottage cheese, start with 1 gallon of milk and add half a rennet tablet to the milk. For faster action you can also add a cup of buttermilk to a gallon of milk, which should be at about 78 to 80 degrees.

Set the milk in a warm place and in a covered container and, after adding the buttermilk and the rennet, allow the milk to sour and then to curdle. When the curds have formed and they float on the whey, which has sunk to the bottom of the container, you are ready to cut the curd.

Use a long, sharp knife to cut the curd. Stick the knife all the way through the curd and to the bottom of the container, and make long, sweeping cuts that are parallel from one side of the container to the other. The cuts should be about 1/2 to 1 inch apart. When you are finished you should have a series of cuts across the entire container.

Now start to cut the other way and make long cuts that dissect the first cuts; these second cuts should also be 1/2 inch to 1 inch apart. When this series of cuts has been made, you will have a container of curd that is chopped into 1/2-inch or 1-inch squares.

Stir the curd well, just enough to keep it open and moving. Place the container on the heat and increase the heat very slowly. As you increase the heat your curd will start to become more and more firm. If you have a candy thermometer or other method of measuring the heat, wait until the cheese has reached a temperature of about 105 degrees, or a little lower but not higher. You can let it reach the body temperature of 98.6 degrees if you prefer. Then hold the heat at that precise point as well as you can and maintain your vigil until the cheese is as firm as you want it.

When the proper state of firmness has been reached, remove the cheese from heat and place the curds and whey into a draining vessel. You can use a sieve or a cloth inside a colander so that all of the firm cheese will be held back while the liquid whey passes through the cloth. You can salt the cheese to suit your taste now and allow it to drain. When it has drained sufficiently, you can store the cheese in your refrigerator until it is time to use it. Be sure that you use it within a week to 10 days, since cheese will not keep very well for long periods, even under refrigeration.

MAKING CHEDDAR CHEESE

Cheddar cheese is rather more difficult to make than cottage cheese, but it can be done if you have the time and the patience. It is a good idea to start with a modest amount of materials and ingredients to see how well you fare.

Start with 1/2 gallon of whole milk and add to it 1/2 cup of buttermilk. Leave this mixture in the heated room—just don't refrigerate or leave outdoors in cold weather—and allow it to "work" for a full day. When you have completed this step, add it to 3 gallons of whole milk in a large container. If you wish, you can reduce the amounts proportionately. Start with a quart of whole milk and add it to 1/4 cup of buttermilk, after 24 hours, add the worked mixture to 2 gallons of milk, if you prefer.

With the milk mixture now at room temperature or a little higher, wait for 2 to 3 hours and then add 1/4 to 1/2 of a rennet tablet, depending on how much milk you used—1/4 tablet for 2 gallons of milk, 1/2 tablet for 3 or 4 gallons of whole milk.

When the rennet has been added to the milk, slowly heat the cheese to about 85 degrees and hold it at that temperature until the cheese thickens properly. At that time, remove it from the heat and prepare it for pressing.

The whey will be involved with the curd, and you don't want to cut it and drain the way you did with cottage cheese. Place the cheese in a tall, open-mouthed container that is a perfect cylinder—a shortening bucket or can, for instance, with the top end cut out and the bottom perforated with holes—using the cheesecloth as a lining for the container.

With the cheese and cheesecloth inside the container and the container itself standing on the perforated end, put a circular weight that just fits inside the container on top of the cheese, but the weight should be in contact with the cheesecloth rather than the actual cheese. The weight can be made of a disc of wood, and small weights can be laid on top of the disc. As the cheese is pressed, the whey will drain out the holes in the bottom of the can, and the weights will slowly compress the cheese into a solid mold.

If you wish to speed up the draining, you can slice the curd and stir it while it is still warming, and you can add salt to taste at this time.

After the cheese is in the press, leave it until it has drained and solidified, and then store it to cure in a cool, dry place—one with an ideal temperature in the mid-40s if possible and in the low fifties at the highest for best results.

The longer the cheese is cured the better the flavor, so plan to leave it in the curing stage for 6 months or so. Three months of curing will give one flavor, and 6 months will produce an entirely different degree of sharpness. Three additional months will add even more depth and fullness of taste.

After the cheese is cured, you can freeze it by wrapping it in butcher paper and storing it without any pretreatment in your freezer. You can grate cheese for use in salads and other dishes and freeze the grated cheese and later use it without thawing. If you wish, you can slice the cheese and put the slices over food to be cooked later and freeze the whole dish.

You can also store the cheese in wax by melting paraffin and dipping the entire mold of cheese into the wax, allowing the wax to cool and seal the cheese from the air-causing spoilage. Be sure that the wax completely covers the cheese and that it is coated sufficiently to prevent spoilage. Refer to the chapter on jams and jellies and use the same general principles that were employed in the sealing of jellies by paraffin—the only difference is that the seal completely envelopes the cheese.

You may notice that, if you make several batches of cheese, no two batches will taste exactly alike. This condition is normal, just as you may find a variety of tastes in a bushel of apples grown on the same tree or the variety you will find in two jars or more of jams or preserves. Cheese flavor depends upon moisture, the type of milk, the time of curing, and the temperature at which the curing occurred.

You will also notice that your cheese is not as deep and rich colored as that you purchase in the grocery store. There is no problem here, either, since the commercial cheese processors and packagers use cheese coloring to achieve the desired coloring. You can even order the cheese coloring from manufacturers if you feel that you need it, but the natural color of the homemade cheese is fine and there is no problem with your work. The taste is the final and most important test.

MAKING YOGURT

A close relative of cheese, particularly cottage cheese, is yogurt, which is milk in which bacteria have been injected. To start your yogurt, you must buy the yogurt starter and then continue to make the yogurt by using a little of the commercial starter in each batch you make at home. Because of the highly complex bacterial elements of yogurt, you cannot, under normal conditions and circumstances, make yogurt starter at home.

Start with regular milk and introduce the bacterial starter, then incubate the yogurt at 110 to 112 degrees for 3 hours. If you don't maintain the proper temperatures, the yogurt will simply become soured milk, unfit for consumption.

When you start your home efforts, the simplest method imaginable is to add 3 tablespoons of commercial yogurt to your own quart of milk. The bacteria, which are capable of killing nearly every form of harmful bacteria within 24 hours but are absolutely harmless and even beneficial to human beings, will start converting the milk into lactic acid. When the yogurt is ready, place it in the refrigerator but do not eat all of it. Save 3 tablespoons of it to

Fig. 14-1. To make yogurt, first pour the milk into a cooking vessel, bring milk to a near boil, then allow to cool to body temperature.

Fig. 14-2. Spoon in the yogurt you bought and stir it until it mixes with the milk and starts to curdle.

Fig. 14-3. Pour the mixture into a flat container and then prepare the dehydrator.

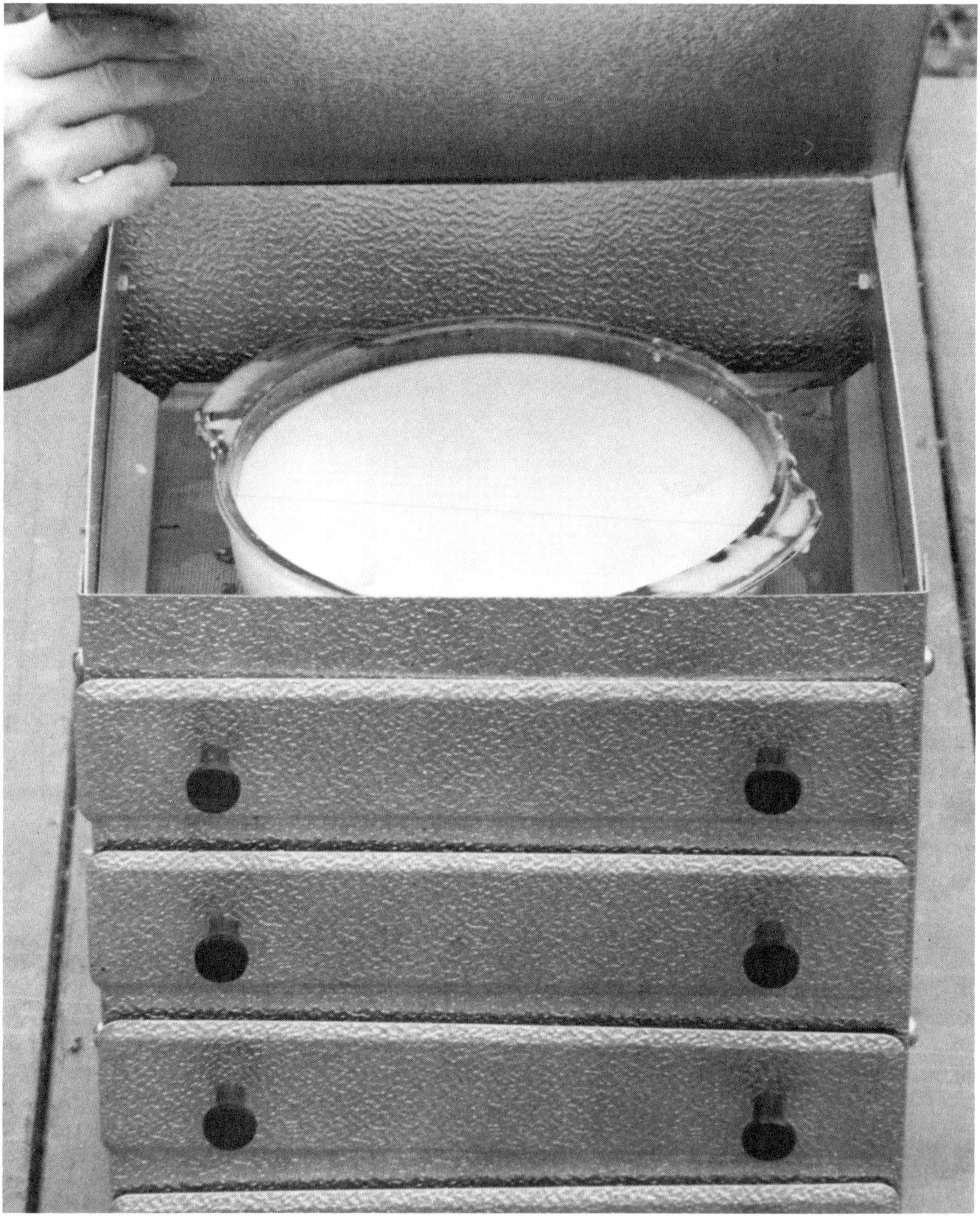

Fig. 14-4. You can lift off the lid of the dehydrator in order to put a large container inside. You can also remove one or more of the racks. Leave the yogurt in the dehydrator until the thickening is completed.

put into your next quart of milk. A quart of milk makes 1 cup of yogurt.

Some expert yogurt makers recommend that you boil your milk and then allow it to cool to body temperature, add the bacteria or tablespoonfuls from a previous batch of home-made yogurt, and maintain the temperatures at 110 to 112 degrees, then refrigerate when the yogurt is ready (Figs. 14-1 to 14-4).

By adding spoonfuls of yogurt to the next quart of milk, you can continue the cycle for a long period of time before the bacteria strain weakens. Every 3 or 4 weeks buy a new supply of yogurt at the grocery store.

15

Preserving Juices

Juices, whether fruit or vegetable, are among the most desirable of preserved foods. The juices, which can be canned or frozen, can be kept as long as canned or frozen foods can be preserved.

In May and June you may have access to a large amount of strawberries, which do not yield much juice. The juice may be extremely tart and hard to obtain. In July blackberries are ripe in many parts of the country. They have much more juice that you can extract and can or freeze.

If you live in an area where citrus fruits are readily and plentifully available, you can make wonderful citrus juices. For the majority of the country, however, the best bets are probably apples and grapes, plus assorted vegetables.

JUICE EXTRACTION

There are 2 major ways of extracting juice from fruits: pressing or cooking.

Pressing Juices

If you choose to press the fruits, you may do the major work by buying or borrowing a fruit press, or you can do it largely by hand.

A fruit press is simply a wooden cylinder with a wooden or metal disc that is pressed down on the fruit by means of a crank or handle that is turned steadily. Cracks in the cylinder wall permit the juice to flow from the cannister and into a trough, which eventually flows to a container below.

The fruit press works very well with grapes. You may need something stronger and sturdier for crushing apples and other very hard fruits. An apple press can be purchased for a nominal sum.

Grape Juice. After you have gathered or bought the fruit, wash it and remove any foreign materials that may have found their way into the fruit (Fig. 15-1).

Put the press on a bench or table that is a good working height, and position it so that you can place a pan or bucket below to catch the juice as it flows from the press (Figs. 15-2 and 15-3). If you do the pressing outdoors, the aroma of crushed grapes will be certain to attract inordinate numbers of

Fig. 15-1. To extract juices by using a fruit press, wash, core, and chop apples into quarters. Use fruit that is spotted or damaged, but cut out the bad spots.

Fig. 15-2. Elevate the press so that you can place a container under it to catch the juices effectively. Use a clean block of wood to push apples down into the press.

Fig. 15-3. Use the handles to the screw for initial pressing.

honeybees, wasps, hornets, and yellow jackets. If you have a screened-in porch, you have the ideal place to work.

Do not try to fill the press too full. It is often much better to fill the cylinder 1/3 full and then lower the metal or wooden disc. Turn the handle steadily and, when it is seated on the surface of the grapes, turn it another 3 or 4 times, then pause for several seconds to let the juice flow start. If you turn the crank too much, you will cause the grapes to split under pressure and juice will squirt in several directions. You want the grapes to split slowly, not explode (Fig. 15-4).

After the juice stops running in a steady stream, apply more pressure, then wait again until the stream slows. Repeat the process, pausing after each 2 or 3 turns of the handle. After a few minutes you will have the disc pressed virtually against the bottom of the press. Turn the handle backward to relieve the pressure and remove the hulls of the pressed grapes. Do not discard the hulls; put them in a large container that you can cover with a piece of plastic or cloth.

Fill the hopper or cylinder again, and press the new fruit. Again, add the hulls to the previous ones. Keep the pattern going. When you have the juice container filled you can transfer the juice to the refrigerator while you press more fruit.

When the hull container is filled, you can empty the press and fill it with the crushed hulls and press them again. You will be surprised how much more juice can be extracted.

Do not throw the hulls or skins away; instead, transfer them to another container after you squeeze

Fig. 15-4. You can use a length of sturdy wood to provide extra leverage if you are pressing firm apples. As you slowly turn the screw, the juice will pour into the container in a gentle but steady stream.

each batch the second time. Run the hulls or skins through the press a third time. On this run you will get far less juice but it will still be enough to make the effort worthwhile.

Now put the hulls in a large pot, pour 1/2 cup of sugar over each quart of hulls, and add a small amount of water—perhaps a cup in the bottom of a large cooking vessel. Put the pot on the burner and turn on medium heat. Cook the hulls for an hour, and at the end of that time put the hulls in a collander and allow the last juice to flow into a container beneath the collander. This last run of juice can be preserved just like the other juice or you can use it to make jelly.

If you elect to make jelly, you can return the juice to a burner. Put the juice in a small cooking vessel and slowly add sugar until the juice starts to thicken until it will form double-drops on a tilted spoonful of juice or jelly.

When the jelly is done, seal it according to the directions given in the chapter on fruit jams and jellies, or leave it unsealed for immediate serving.

Apple Juice. If you wish to extract juice from apples by crushing, you will need a cider press rather than the grape press. If you don't own one, it might be easier to take your apples to a cider mill operation and have the operators press the apples for a small fee. You can buy the press or mill if you intend to stay in the apple business or if you have ready access to large amounts of apples for a long period of time.

You can use any apples, but it is a general rule to relegate to the cider heap all those apples which have fallen from the tree and have been bruised. You can use any which have spots in them, but be sure to cut the spots out before you crush the apples.

After you have extracted all the juice you can from the apples, you can cook the peels and pulp and get still more juice. Again, this juice can be used for drinking or you can cook it for jelly.

Berry Juices. If you are extracting juices from blackberries or similar fruit, you have no real problem. You can either cook the berries slowly and they will instantly yield their juice, or you can wash the berries and then make a sack-like container from a clean cloth and drop the berries into the cloth. Gather up the loose ends of the cloth and tie them together or have someone hold the top of the cloth together. Use both hands and squeeze the bag until the berries are crushed, or you can lay the sack or cloth on the bottom of a container and then crush the berries with the heel of your hand or with a piece of clean 2-by-4.

Cooking Juices

What if you have a moderate supply of grapes or muscadines but have no press and it isn't worthwhile to take the grapes to someone who can press them for you?

For one solution you can cook the fruit after you crush it by hand or by using a small wooden block covered with a clean cloth.

You can use your hands and simply grasp a handful of grapes and squeeze, but you should keep your hands well down in the container so that grape pulp and juice will not squirt over the room and onto your clothing.

If you choose to try the wooden block, get a short piece—1 foot or even shorter—of clean 2-by-4. Do not get a block of wood that has been in a stack of old lumber and may be infested with various germs and insects. If you have a new length of wood, use it; if not, use one that is clean and free of all discernible problems. You can boil the wood for several minutes in a cooking vessel, and you should cover the end of the block with a clean cloth folded several thicknesses when you start to crush the grapes.

Put the grapes into a very strong container. Do not use a plastic vessel that is weak enough that you can punch a hole in the bottom of it with the wood.

Put an inch or two of grapes in the container and then use the wrapped end of the wood to crush the grapes by applying firm pressure on the the wood. Try to make sure that the skin of every grape is broken.

Put the grapes into a cooking vessel and put the pot over low heat. Let the grapes cook slowly while you crush more grapes to add to the pot. Turn up the heat when you have enough grapes to make sure that the pulp will not be scorched.

When all of the grapes are cooked until the juice has been freed, you can put the hulls and pulp through a collander and let all of the juice drain out. Let the pulp drain for an hour or more, and stir the pulp around while the draining is taking place.

Finally, use the pulp for fruit preserves by extracting the seeds and making fruit dishes as described in the chapter on jams and jellies.

CANNING JUICES

You can preserve juices by canning in the same way that you canned fresh fruit. After you have extracted the juice, run it through a jelly bag and then return the juice to a pot and add 1/2 cup of sugar for each gallon of juice. You can omit the sugar if you prefer unsweetened juice. Cook the juice until it reaches the simmering point and then pour it into containers and fill to within 1/2 inch of the top. Close the jars with lids and rings soaked in very hot water and tighten firmly. Process the juice in a water-bath canner for about 5 minutes.

FREEZING JUICES

Freezing juices is extremely simple. After the juice has been extracted, strain the juice to remove all impurities and then add 1/2 cup of sugar to 1 gallon of juice or omit the sugar completely. Freeze the juice by pouring it into a freezer container—you can use freezer bags or plastic cartons. Close the containers well and store the juice in the freezer.

When you use frozen juice, particularly grape juice, you should allow the juice to thaw completely and then stir it well before serving. If you allow it to thaw only partially, the juice will separate. The part containing the greatest amount of sugar separates first and leaves the watery part still frozen. So the first glassfuls of juice are truly indescribably delicious but the final glassfuls are watery tasting and flat.

We advocate freezing or canning the juice without adding sugar. You can always add the sugar, if you desire it, just before serving. On the other hand, if you add it before canning or freezing, you have a great deal of sugar, if you have a huge supply of juice, tied up. If you should later decide to make jelly, there may be some question of how much more sugar should be added.

16

Making Your Own Wine

One of the oldest and most popular forms of home food preservation is the art of wine-making. A great deal has been written and said about how delicate wine-making is and how extremely careful and knowledgeable you must be to help nature to convert sugar in a liquid into alcohol.

The truth is that if there happens to be a puddle of water under a grape vine and ripened grapes fall into the puddle, a crude form of wine may well result from the mixture of water, grape juice, natural yeast, sugar, and temperature.

The resulting concoction naturally will not be a potable wine. You can, however, by exerting a little energy, patience, and a little more understanding, make your own wines at home.

You can make wine from apples, pears, peaches, plums, tomatoes, pumpkins, and any of dozens of other fruits, berries, and vegetables.

LEGAL REQUIREMENTS

According to the law, you may make wine for your own personal use; however you are restricted in several ways. You cannot sell the wine, trade it, or in any other way exchange it for values. You are also limited in the amount that you may make.

You also cannot make wine legally from any sort of grain. "Corn likker" is not legal, but you can make a fine wine from corn cobs.

Since 1933 when the 18th Amendment was repealed, most states permit you to make up to 200 gallons a year for personal family consumption, but not to sell, give away, or trade. The law further stipulates that the wine maker must be the head of the household, and that anyone wishing to make his own wine must fill out Form No. 1541 and send it to the Alcohol Tax Unit of the local office of the Internal Revenue Service.

Check with local authorities to see just how the law is worded in your own state or county, since some states may require home licenses.

EQUIPMENT NEEDED

Actually, depending on how extensively you plan

to get involved with wine-making at home, the equipment needed may be very modest indeed, or it can be very elaborate and also costly.

For the cheapest possible methods of making wine, you need to obtain only a large container—preferably of thick glass or plastic—a sheet of plastic, and a 4-foot length of plastic hose about 1/4 inch or less in diameter.

In fact, if you want to be even more economical, you need to have only a gallon plastic or glass jug, a balloon and a piece of ordinary twine or cord of the type used to close feed sacks or fly kites.

If you want to be a little more sophisticated you can buy a hydrometer, fermentation lock, and plastic hose the size of a drinking straw but several feet long.

You can buy all of the expensive equipment, including charcoaled oak casks or barrels and similar luxuries, if you prefer. There are some optional items you may also want to have on hand, such as sodium bisulfite tablets, yeast nutrient tablets, and wine-making yeast.

PREPARING FRUIT

First of all, as you would in any home food preparation process, choose fruits that are ripe, firm, and sound. Do not attempt to use fruits or vegetables that are overripe to the point of softness or underripe to the point of hardness. Do not use any fruit or vegetables that have blight spots or rot spots on them. Wash all fruits and make certain that there are no foreign matters present as you get started on the preparation of the fruit.

Above all, wash carefully enough that you will get rid of all insecticides, in addition to the dirt and wild yeast that is present in nearly all fruit.

If the fruit is of a high-acid content, particularly the skins, you may want to peel the fruit. In other cases you will want to leave the skins, peels, or hulls in during the first stage of the wine-making. Orange and grapefruit skins, for instance, have too much acid to be left in the process, but the skins of apples and most grapes are acceptable and even desirable.

If you are planning to use apples, you can core and slice the fruit. Pit and peel peaches; simply crush grapes; wash blackberries and crush either by hand or with a length of 2-by-4 wood.

If you are using dried fruit—and dried fruit can be used to make excellent wines—you should drop the fruit into boiling water and let them soak as the water cools so the fruit will be returned to its former size and consistency.

You can use some commercially prepared fruit juices and fruit juice concentrates, but be sure to buy those items that have no preservatives, such as benzoate of soda, in them.

After your fruit has been crushed and otherwise made ready for processing, you should determine the sugar content of the juice. Some grapes and other fruits are so sweet that they do not need to have sugar added; others must have sugar mixed with the juices. The presence of sugar is not simply a question of whether you want sweet or dry (unsweetened) wine; the sugar must be present for the conversion to alcohol to occur.

You may make a wine with an alcohol rating of 14 percent, which is the greatest alcoholic content possible under all normal conditions in wine-making. You certainly can have alcohol levels much lower, however, but when the content drops below 12 percent, there is a definite danger of spoilage. If there is too much sugar, there might be an incomplete or stuck fermentation. Obviously, then, the sugar level must be fairly close to the precise point.

If all conditions are right, wine may be kept for years. On the whole, however, any wine kept more than 5 years is likely to be highly acidic or vinegary.

If the juice is excessively acidic, as is the case of Concord and Niagara grapes, you will probably need to add 50 percent as much water as you have juice. If you have 2 gallons of juice, add 1 gallon of water before you make the sugar test.

The sugar percentage of juice is known as the **Brix level** or content, and the ideal level is 24 percent. If you have a hydrometer, fill the cylinder about 3/4 full of juice and then lower the hydrometer with the heavy end down into the cylinder and allow it to settle. Note where the juice comes in relation to the calibrated marks on the sides of the hydrometer.

Subtract the Brix measurement from 24, which is the ideal level. For each percent difference on the scale, add 1/4 cup of sugar. For example, if the Brix level is 20 percent, you are 4 percent marks off the ideal level. Add 1 full cup of granulated sugar to the juice for each gallon of juice. All measurements are in terms of gallons. If you have 2 gallons of juice and the Brix is 4 percent off, you need to add 2 full cups of sugar.

MAKING AND BOTTLING THE WINE

After the Brix mixture is correct, pour the juice with the sugar already stirred into it into a container or the fermentation tank. Add 1 yeast nutrient tablet, which should be 8-grain tablets, and see that the nutrient is mixed well. One easy way of mixing is to crush a tablet and then pour the crushed tablet into a small glass half full of juice. Stir completely and then return the juice to the fermentation tank. You can also mix added sugar in the same fashion.

You can also add a tablet of sterilizer (sodium bisulfite) to prevent the wine from spoiling while it ferments and after bottling.

Now empty 1 packet of wine yeast into a small cup of lukewarm water and wait for the yeast to become activated. After 5 to 15 minutes you can pour the yeast into the container and install the fermentation lock.

When you have all of the ingredients in the tank or bottle, you must close the container so that it is air-tight, but, **and this is crucial,** the wine as it ferments will build up tremendous pressure, enough to shatter the glass bottle or split open the plastic container, unless you allow the carbon dioxide and other gases to escape.

This is where the fermentation lock comes into play. The lock is a simple plastic device that has a vent tube which will allow gas under pressure to be forced from the tank or container. As the gas leaves the tube, it is forced then into a tiny container of water. The gas forms a bubble which passes through the water; then the gas escapes into the air.

Because of the water in the fermentation lock the air in the room cannot pass into the tube and then into the wine. With the fermentation lock, then, you have created a one-way street; gas can escape but air from outside cannot get into the tank (Figs. 16-1 and 16-2).

You can let the tank sit and bubble for as long

Fig. 16-1. Five-gallon jars with fermentation locks will serve to make large amounts of wine. The lock at the top allows gas to escape but will not permit air to enter the wine area.

Fig. 16-2. A simple but effective fermentation lock can be made by using a cork and a length of plastic tube, plus a bottle of virtually any sort. Run the plastic tube from the cork to the bottle, which is filled with water. Nature takes care of the rest.

as the fermentation continues. You need not take any special precautions other than taking care not to allow the temperatures to rise too high. In other words, keep the fermenting wine at room temperatures.

As long as the bubbles rise, the wine is still working. After several days—the time varies considerably from 1 type of wine to another—the bubbles will cease. Let the wine set for several more days. Watch the wine carefully to see that the working has indeed stopped. If you should bottle wine that is still working, the pressure will continue to build until either the cap or cork will be blown from the bottle or the bottle itself will explode with enough force to cause serious injury to anyone in the immediate area. We have had bottles explode and make the same noise as a shotgun blast. The result has been glass fragments embedded in the paneling of the walls and a hole in the ceiling.

When the fermentation is completely ended, there will be a thin substance that will appear gradually on the bottom of the fermentation tank. This is, in essence, a type of tartar or dead yeast, and you do not want this to get into your bottled wine.

When you bottle the wine, you can use a siphon tube or small plastic hose and fill it with water by holding your finger over one end and allowing the tube to fill with the tiny stream from the kitchen faucet. You can turn the tube upside down and the water will not pour out as long as you have your finger tightly over the other end.

Now, with the fermentation lock off the tank, lower the end of the siphon tube into the wine, then lower your hand with the finger over the end until the outside end is lower than the end in the fermentation tank.

Hold the outside end over the opening of a sterilized bottle and quickly shove the end into the bottle immediately after you remove your finger from the end of the tube. The tiny amount of water will pour into the bottle and the water will be followed instantly by the wine.

Take care not to let the end inside the tank get low enough to allow dead yeast into the bottle. Keep the end 2 or 3 inches off the bottom and let the wine fill the bottle except for the last inch or so of head room. When the bottle is full, pinch the tube to cut off the wine flow and change the end from the bottle you have just filled to the next sterilized bottle, then proceed to fill it and as many others as you have available as long as the wine supply lasts.

When all the bottles are filled, cap or cork them instantly. Cork, incidentally, works better (Figs. 16-3 and 16-4). Wine needs to "breathe" and air can actually filter in incredibly small amounts—not enough to allow spoilage—from the wine to the outside. Screw caps will not permit this breathing, but if you have no corks you can use the caps.

ALCOHOL CONTENT

If you intend to make much wine, you should purchase an inexpensive hydrometer—these can be bought at numerous wine supply houses or lab supply stores—so that you can take accurate alcohol content readings. You want a healthy alcohol content in your wines to lessen spoilage likelihood and to get a better taste in the finished product. Since watermelon wine is so unusually variable, it would be wise to start here with the testing.

The test can be made in one of two or three ways. The first and simplest way is to take a hydrometer reading of the "must," the wine mixture before fermentation begins. After the wine is ready to start fermenting, pour a small amount of the mixture into the hydrometer and record the reading that you find on the "Potential Alcohol by Volume" scale. If your must reading, for instance, is 2.4 your potential alcohol level is 16 percent. After the fermentation has been completed, take another reading. In the second reading, you may find, for instance, that the level is 4 percent. Subtract the second reading from the first reading, and you learn that you have an alcoholic content of 12 percent for your wine.

A second method, also involving the hydrometer, is to pour exactly a pint of fermented wine into a container and use a part of the wine to make your hydrometer reading. Record the reading, return the wine to the pint container, and pour all of the wine into a glass cooking vessel or

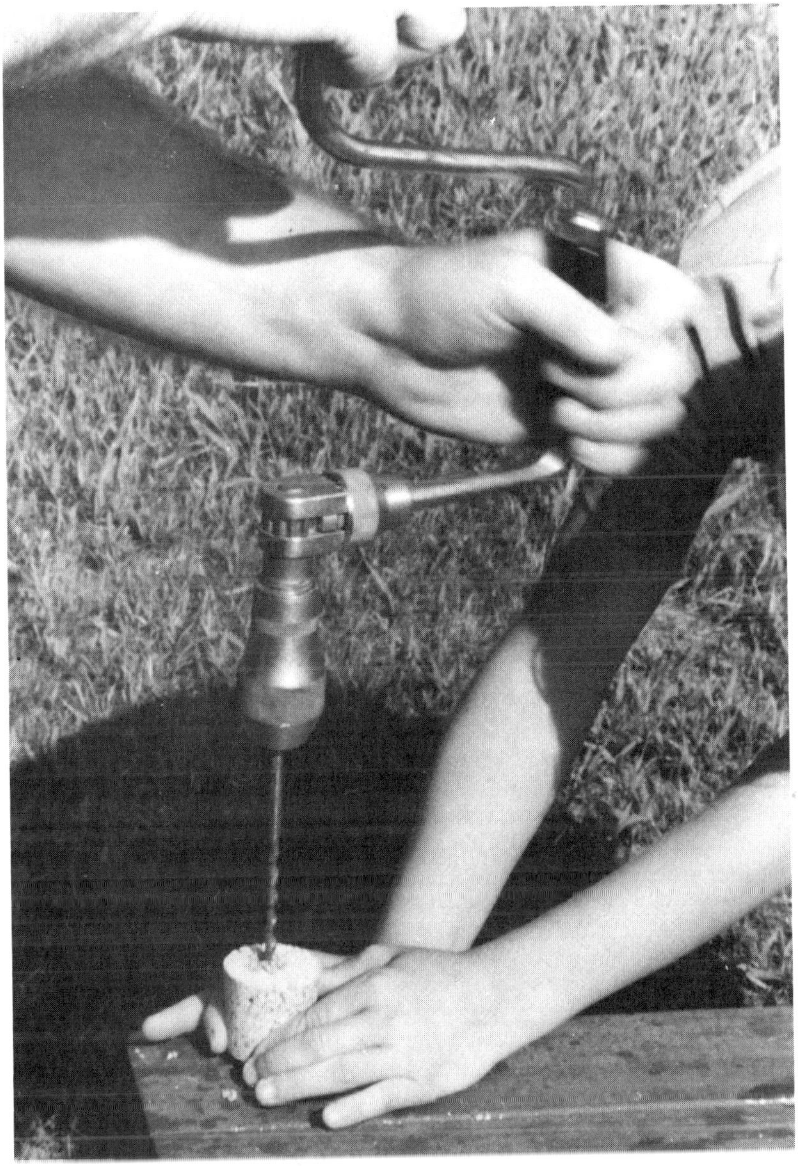

Fig. 16-3. You do not need a special cork for making wine. You can buy any traditional cork and, using a small bit, bore a hole straight through.

enameled pot. Boil the wine until only 1/2 pint is left. Make a second reading. The difference between the 2 readings is your alcohol content.

The principle involved is that boiling will kill the alcohol, since alcohol boils at a much lower temperature than water. When there is only half a pint left, the remaining liquid will be the "wine," less all the alcohol. If your difference between the 2 readings is 10, your alcohol content is about 7 percent, if the difference there is 11, the alcohol content is 8 percent. If the difference is 18, the alcohol content is 14 percent.

You can make a fairly accurate estimate of alcohol content based on these figures. For instance, if the difference between the 2 readings is 14, and since you know that a 10 will give you 7 percent and 18 will give 14 percent, you can assume that the reading of 14, which falls roughly halfway

Fig. 16-4. After you have bored a hole in the cork, insert the end of a plastic tube. If the hole is too large, wrap masking tape around the end of the tube to create an air-tight closure.

between 10 and 18, will yield an alcohol content reading also roughly halfway between the 2 alcohol readings of 7 percent and 14 percent.

While this reading is not at all accurate in the scientific sense, it will give you a rough idea of how much alcohol is in the wine.

The difference between 7 percent and 14 percent means that, if you were to write down the following figures—7, 8, 9, 10, 11, 12, 13, 14—you have represented the 2 extremes in the test. Since you are working with test figures halfway between these 2 points, you should mark a point halfway between the 2 outside figures. The result would be exactly halfway between 10 and 11, so your alcohol content would be 10.5.

If this sounds complicated, remember that your hydrometer will show you exactly what figures apply to your particular measurements. The major advantage in using the hydrometer is that it will show you what wines are more likely to keep well during the aging processes and which ones are more likely to spoil because of low alcohol content.

If you have a wine with very low alcohol content—5 to 7 percent—you may wish to increase the alcohol volume content to reduce the risk of spoilage.

To do so, use a siphon tube to drain off all wine and leave the sediment, pulp, dead yeast, and any other undesirable elements in the carboy or large container that held the original wine must. Put the siphoned wine into a clean container and then start a new cupful of yeast by sprinkling a small amount of yeast into the warm water. Allow it to stand for 5 to 10 minutes.

You can now heat 1 1/2 pints of water in a saucepan and mix 2 cups of sugar. Stir until the sugar is totally dissolved. Finally, pour the sugar water and yeast into the wine and attach a fermentation lock. A new fermentation process will start within 24 hours and a stronger wine will result.

You can run the wine more than twice if you desire, but when the alcohol level reaches 14 percent the yeast cells tend to die quickly; so it is not likely, unless you fortify the wine by adding brandy or another high-alcohol beverage, that you will achieve a higher alcoholic content.

THE EASY AND INEXPENSIVE WAY TO MAKE WINE

If you do not have the time, money, or inclination to devote a great deal of yourself to making a jug of wine, you can make a reasonably good wine with little or no cost in equipment and very little in ingredients. Start with about 8 pounds of berries, if fresh, or 10 pounds of canned but drained berries. Keep the juice after you drain the berries if you used canned ones.

Crush the berries with your hands or with the bottom of a thick glass, and mix 4 pounds of sugar in with the berries. Stir well. Now add a packet of yeast that has been dissolved in warm water. If you have campden tablets and yeast nutrient, crush one campden tablet and add 1 teaspoon of yeast nutrient. Finally, add enough water, juice, or a mixture of both, enough to make, counting berries, 2 gallons of wine starter or mixture.

At this point you are ready to pour the juice and berry mixture into two 1-gallon plastic milk jugs that have been cleaned and sterilized. You should wash them in very hot water or wash in hot water and pour boiling water in the jugs, sloshing it around to kill as many unwanted microorganisms as possible.

After the jugs have been filled—try to divide the berries and juice equally between the two jugs and leave an inch or so headspace in each jug—stretch a fairly large balloon, one that can inflate up to the size of a large head of cabbage, over the opening of the jug. Set the jugs aside and watch the action.

As soon as the fermentation starts, the balloon will start to inflate. Soon the balloon will be inflated to near-capacity; in fact, it may burst; so be sure to keep a supply of extra ones on hand.

After several days the balloon will reach its peak of expansion and will remain that way for several days. Then, slowly, the balloon will start to deflate. Watch it daily as it becomes smaller, and when it is totally flat, remove it and bottle the wine.

If you can pick your own berries, the total cost of the 2 gallons of wine will be about $1.50, depending on how much you must pay for sugar and yeast.

Let the wine age for 2 or 3 months before you sample it. If you use corks for stoppers, lay the bottles on their sides so that the wine can stay in constant contact with the corks, and turn the bottles 1/2 revolution every week or so.

SOME BASIC WINE RECIPES

In virtually all wines, if you have the ingredients, use one campden tablet and 1 teaspoon of yeast nutrient, plus 1 packet of yeast for 1 to 5 gallons of wine. For each gallon of wine you will need about 2 pounds of sugar for most fruits that are prepared fresh. For a few very tart or acid fruits, such as cranberries, you can use 3 pounds of sugar. Fruits and berries that can be made from this basic recipe are blackberries, elderberries, apples, cherries, apricots, strawberries, grapes, loganberries, scuppernongs, muscadines, peaches, plums, and raspberries.

Banana Wine

To make banana wine, use fresh or dried bananas. If the bananas are fresh, peel, mash with a fork until they are almost liquid, and set aside until later. Use about six or seven large bananas.

If you use dried bananas, use 1/2 pound. Add 1 1/2 cups of raisins. In a separate container put slightly less than 1 gallon of warm water. Add 2 pounds of sugar. Stir until the sugar is dissolved and mix the yeast that has been dissolving into the water. Add the bananas and raisins. Drop the raisins into hot water and let them plump before you add them to the mixture.

After you add the yeast nutrient and campden tablet—if you don't have them the wine will still ferment and be very palatable—put on the fermentation lock and let the wine ferment. After about 10 days pour off the wine and drain the raisins and other ingredients. Don't discard the water and wine mixture that drains. You may discard the raisins and any other solids and wash out the container. Now pour the water-wine back into the container, reattach the fermentation lock, and let the wine continue to ferment. If you prefer, you can leave the raisins and bananas in until the fermentation has stopped, then drain off the wine when you plan to bottle it.

Other Wines

Dandelion wine can be made by picking 6 to 7 cups of dandelion leaves—do not use the blossoms. Mix it with 2 cups raisins, 1 gallon, or slightly less, of water, 2 cups of granulated sugar, a campden tablet, yeast, and yeast nutrient. Let it ferment for a week in the container with the fermentation lock attached, then drain off the wine and discard the other ingredients, or you can leave the entire mixture together until the fermentation has ceased completely. Age this wine, as well as blackberry wine, for at least 3 months.

Rhubarb makes a fairly good wine if you chop 3 cups of rhubarb finely, mix 2 1/4 cups of sugar over it, and pour in 1 gallon of warm water. Mix thoroughly. Add the yeast and other ingredients listed above and put the mixture into the fermentation tank and let it work until fermentation ceases, then pour off the wine into clean, sterile bottles, and cork. Age the wine for 3 months, longer if you can wait.

Wine From Concentrates

You can buy the fruit concentrate if you prefer not to squeeze berries and other fruits. To make wine with any of the wide variety of concentrates, mix 1 can of concentrate to 7 cans of warm water. If you use a 3-quart can of concentrate, use 15 to 16 pounds of sugar, 4 campden tablets, 2 teaspoons of yeast nutrient, 1 packet of wine yeast, and a handful of raisins, if you prefer. Mix all but the yeast and let the sugar and other ingredients start to work. After the mixture is cool, you can add the yeast. Never add the yeast to a hot or very warm mixture.

Pour the mixture into the fermentation tank and attach the lock. Let it work until fermentation ceases and then wait another 3 to 4 days. Bottle and cork. Let the wine age 3 months.

MAKING A MUST

Many people who make wines insist that starting

by making a must is, indeed, a must. They point out that if you mix your grapes or other fruit, add the sugar, yeast, and other ingredients, then put all of the mixture into a large plastic container and cover it with a sheet of plastic, you will have a better wine. This early mixture is called a **must.**

The plastic sheet must be virtually air-tight, and you will find on the next day's inspection that there is a crust of wine-fruit and hulls or skins packed at the top of the container.

Use a long-handled spoon to break the crust and stir the must well. This must is sometimes called a primary fermentor. Leave the fruit and water in the must for 6 to 7 days. Then drain off the wine-water and place it in a fermentation tank and attach the fermentation lock.

Let the wine proceed to work, and when fermentation ceases, bottle and age the wine.

Some people also keep the must ingredients and add more water and sugar and make a second-run wine, which is generally a little lacking in color but otherwise is a good wine.

We have tried making wine both ways, with a must and without, and we can't really tell a great deal of difference.

STUCK FERMENTATION

Occasionally there will be a slight problem, called *stuck fermentation,* in which the wine will start to ferment and then, inexplicably, the working will stop. In such an instance, you must make necessary corrections or the wine will never be worth drinking.

To correct stuck fermentation, make a few minor adjustments. You may have the wine temperature too low, and since many yeasts will not work well at temperatures below 65 degrees, if you move the wine to a warmer room the problem may correct itself.

If you are making wine in extremely hot weather and the wine gets warmer than 85 degrees, your yeast action may have been killed. Mix another packet of yeast and stir it into the wine, after moving the wine to a cooler room.

You may also try to siphon off a small amount of wine, add to it some new yeast and nutrient, and stir in a cupful or more of sugar and wait until the small wine container shows signs of fermentation. Return the wine and the new sugar and yeast to the original container and the fermentation may proceed as normal.

SWEETNESS AND AGING

While many people like the dry wines, at least as many are partial to the sweeter wines. What sort of compromise should be made if the finished wine is to be acceptable to all concerned?

The answer is none. You can make a dry wine, age it, and, when you are ready to serve the wine, you can open one bottle and serve it as it is. Open another bottle and pour about 3 ounces into a glass sweeten with 2 tablespoons of granulated sugar or more, depending on the taste of those who will drink it. Pour the small glassful back into the full bottle and allow the sweetened wine to permeate the remainder of the bottle. You will then have 1 bottle of sweet wine and 1 bottle of dry wine.

How long must wine be aged? The answer is again a compatible one. You can actually drink wine before it has even finished fermenting, or you can age it 2 or 3 years. You can drink wine after it has been on the aging shelf for 2 days, 2 months, or, for that matter, only 2 hours.

Wine does, however, improve in flavor and bouquet, or aroma, significantly with aging, and usually it should age at least 9 months but never for 25 to 50 years, however, like the wines you always read about and expect it to be really superior. In virtually every instance involving homemade wine you will have a vinegar that isn't really worthy of being a cooking wine.

You can generally plan to age wine up to 2 years for the best possible taste. One sensible idea is to make a batch or run large enough for you to store several bottles. You can open 1 at 6 months or less, and open another a month or so later. By the time you get to the final bottles you will have decided what the best time, for your personal taste, for aging is.

MAKING COLD DUCK

When you are in the store, you notice that bottles of cold duck are rather expensive. You will perhaps be told that the wine must be prepared commercially so that bubbles can be pumped into it to give it the fizz and sparkle. You may also be told that cold duck is not really a good wine, that it is too weak, too sweet, or lacking in other qualities.

Factually, you can easily make cold duck at your home, and it takes very little working time and very little trouble. You can make a cold duck as dry or as sweet as you like it, and you can in almost every reasonable instance make cold duck from wine that you have purchased.

To dispel another misconception, you do not have to make cold duck or any other type of sparkling wine from grapes; you can make it from tomatoes, pumpkin, sweet potatoes, or whatever else you desire.

In its simplest sense, cold duck is merely wine that has been fermented twice, once in the fermentation tank and once in the bottle. After the wine has been fermented once and bottled and corked, you can open a bottle and start a second, in-bottle, fermentation. It is better if the wine has been aged for 3 to 6 months.

Open the bottle and pour 2 ounces into a glass. You may as well do this with several bottles so you can have several sparkling wines aging.

Start a good wine yeast dissolving, and, when all is ready, return to the small glassful of wine you have poured. Put into the glass 1/2 teaspoon of sugar and about the same amount of yeast; then stir, pour the wine back into the bottle, and cork tightly. Instead of corks you should use the plastic stoppers that come in cold duck and champagne bottles.

One highly important point: Do not put the wine into the usual wine bottles. Use champagne or cold duck bottles, since these bottles are made extra thick to withstand the pressure of the fermentation in the bottle. Keep in mind that cold duck produces fermenting pressure anywhere from about 30 pounds to 90 pounds per square inch. An ordinary wine bottle would explode into fragments as the expansion occurs.

Repeat the process with each bottle you open. As the wine ferments a second time, the carbon dioxide that caused the bubbles in ordinary wine-making will not be able to escape. It is this trapped carbon dioxide that will cause the pop and the sparkle, the effervescence that is so characteristic of these wines.

The dead yeast cells will accumulate on the bottom of the bottle in many instances. If you are not careful, you will pour the yeast into the glasses when you open the bottle, because the effervescence causes the yeast cells to float around wildly in the bottle.

You can avoid the unpleasant yeasty taste and still keep the sparkle when you transfer the wine from 1 bottle to another as you remove the yeast cells.

When the wine has been on the shelf for 6 months or so, place the bottle in the freezer and leave it until thin slush starts to form. The yeast cells will freeze tightly to the bottom of the bottle, and the ice will encapsulate the carbon dioxide molecules. When you open the bottle, the slush will pour easily from 1 bottle into the funnel and into the serving bottle. The yeast remains behind. Cap or stopper the second bottle, move it to the refrigerator area, and leave it until time to serve it.

SPECIAL TYPES OF WINE

It was mentioned earlier that you can make wines from nearly everything that grows. The fruits and berries are obvious, but there are many other ingredients that can be used.

You can, for instance, make wine from tomatoes, corncobs, beets, watermelons, cantaloupes, honeydew melons, celery, turnips, and many other vegetables and other garden favorites.

Corncob Wine

To make corncob wine, use the cobs after you have cut off the kernels. Use several corncobs, preferably those from the juiciest corn you can find. Break or cut the cobs into 3 or 4-inch sections and place these sections in a pan. Add enough water to half cover the cobs.

Turn on the heat and boil the cobs vigorously for several minutes; then let the panful of cobs set for several minutes, until the water has cooled. Take the juice that has been cooked from the cobs and add enough water to make enough liquid for 2 bottles of wine. Add 2 pounds of sugar, the yeast, yeast nutrient, campden tablet, and either the juice of 1 lemon or a small can of grape or other juice. Put the mixture into the fermentation tank and proceed as you would for any other type of wine.

Tomato Wine

To make wine from tomatoes or other vegetables, chop up the tomatoes, add enough water to cook them, and boil tomatoes and water together as if you were planning to can tomatoes. When you are through with the cooking, let the water cool and put the juice into a sieve to strain out all the sediment and seeds. Proceed as you did in the above recipe, except that you would use the juice from a lemon rather than the grape juice. You can even omit the lemon juice if you prefer.

Do the same with other vegetables that you would like to experiment with. Chop the vegetables, boil, then proceed by using the juice extracted and add enough water to make the desired amount of wine. You can also use the basic recipe given above to make wines with several types of dried fruits.

As stated earlier, do not neglect your vegetable garden as a source of wine-making foods. Tomatoes, oddly enough, produce a white wine, not a red one; in fact, nearly all fruits and vegetables produce wines that range in color from white to light shades of pink or near-amber. If you want a red wine, you will need to use the grapes plus the skins, peels, or hulls, since the dark color comes from the hulls, not the flesh inside.

Watermelon Wine

In most vegetable gardens there is likely to be one corner devoted to the growing of watermelons, the summer's universal favorite. If you have only a few melons, by all means eat them while they are at their flavorful best. You will probably wish to dehydrate some for wintertime snacks.

If you have access to a surplus of watermelons, or if you have one that is not really very tasty, you may decide to make watermelon wine.

When you start, do not think in terms of country music or Southern nostalgia. Think rather in terms of a rather fragile pink wine that can be tricky to make and even disastrous unless you pay attention.

If you have a melon that is large, up to 35 or 40 pounds, you will need only part of it to make 5 gallons of wine; so you can eat the best part of the melon and use the ends for the wine.

You will need about 10 pounds to 12 pounds of melon flesh for the wine. Use a table knife or spoon to gouge or cut the flesh from the rind. Dump it all into a huge cooking vessel or plastic container. When you have all the melon cut from the rind, you can start extracting juice.

The simplest way is to get handfuls of melon and squeeze the juice from the flesh simply by tightening your hands together and letting juice and seeds run back into the pot. Keep squeezing until there are no large solids left; you should have a potful of pink juice and seeds (Fig. 16-5).

You cannot really ever expect to squeeze out all the juice; so after you strain the juice to get rid of the seeds and pulp, notice how much juice you have. If you have well over a gallon but not 2 gallons, add more melon flesh and squeeze until you have at least 2 gallons of juice.

At this point you can put a small amount of yeast—no more than 1/4 of a packet of commercial yeast—in a cupful of warm water and let the yeast begin to activate.

Now put 5 pounds of sugar into a large pot and heat 2 quarts of water to pour over the sugar. Be sure that the water is hot enough to dissolve the sugar completely.

Add enough cold water to bring the temperature of the sugar water to a lukewarm stage and pour the sugar mixture into a carboy or 5-gallon tank with the watermelon juice already in it (Fig. 16-6). Finally add the watermelon pulp—and don't worry if seeds get into the wine—and add enough water to bring the mixture to within 5 inches of the top of the container.

Do not fill the carboy all the way to the top. If you do the melon pulp will swell and eventually begin to float. It will either force the cork and

Fig. 16-5. After you have squeezed the watermelon flesh, you can extract more juice by squeezing the sections of rind.

fermentation lock out of the container and your wine will spoil because air has gotten into it, or it will force pulp into the fermentation lock and, since no air can escape, the carboy might shatter.

Many wines will take 3 or 4 days to start working or fermenting, and when they do start they will work at a very slow rate. If you use a plastic tube and water-filled bottle as a fermentation lock, you can time the working by counting the number of air bubbles that rise to the top of the water in 1 minute. It is likely that you will have no fewer than 20 and no more than 35 or 40 in 1 minute.

With watermelon wine, however, the fermentation process becomes highly active in a matter of 5 or 6 hours, and it is not at all unusual to have as many as 300 or 400 bubbles in 1 minute. In fact, the bubbling is so rapid that it becomes impossible to count the bubbles as they rise.

Making Cantaloupe Wine

You can use the same basic approach to making cantaloupe wine as you did for watermelon wine, but you may prefer to make small amounts, since the potential problems are even greater with cantaloupe wine-making.

If you want to make a gallon of this wine, start with 1 small to medium cantaloupe. Cut it in half; then slice it into 1-inch slices. Cut the meat from the rind. Next cut the slices into 1-inch lengths and squeeze them with your hands or press them through a sieve to reduce them to pulp. By the time you are through, you should have a small pot of cantaloupe pulp and juice (Fig. 16-7).

Heat a small amount of water so that you will have about 2 pints of hot water in which to mix 3 cups of sugar. Stir until the sugar is completely dissolved.

Fig. 16-6. When the sugar is completely dissolved, pour the sugar water into the fermentation container and then attach the fermentation lock.

Fig. 16-7. After the fruit or flesh has been chopped into cubes, you can press it into pulp and then allow the juice to strain through a sieve.

Fig. 16-8. Leave some room at the top for the wine to "work." If you do not, the increased pressure can either cause the cork to pop from the jug, or, if air cannot escape fast enough, the jug may explode.

Start a small amount of yeast in a cup of warm water, mix all ingredients and pour into a gallon jug. Fill the jug to within 5 inches from the top (Fig. 16-8). Do not overfill. Attach the fermentation locks and set the jug in a room with a moderate temperature—70 degrees is fine.

You will be astonished to notice that the activity in the jug starts almost immediately. In fact, within an hour you will probably note that the pulp is floating and pushing toward the top of the jug, despite the fact that you left 5 inches of space. The bubbling will begin, too, within an hour or so if all conditions are ideal.

If you overfilled the jug, pulp will work into the fermentation lock and close off the vent which allows the gas to escape, and you may have cantaloupe pulp flying across the room.

Like the watermelon wine, cantaloupe wine will work vigorously for 2 or 3 days and then the working will slow dramatically. Within a week it will be very normal, as compared to grape wine working.

WINE FROM VEGETABLES

You will recall that wine can be made from virtually any vegetable. Successful wines have been derived from carrots, celery, cucumbers, turnips, and a wide variety of melons. Your level of success will depend on an immense variety of circumstances and variables, but you can have considerable enjoyment in the experiments.

If you desire to make wine from zucchini, chop up the zucchini into small cubes and boil the vegetable vigorously in water. One gallon of water mixed with juice will be necessary for a gallon of wine. Pour the juice/water that has been tested for Brix reading and has had the sugar added into the fermenting tank or jug, add the yeast, attach the fermentation lock, and proceed as you would with any other wine.

Do the same with all vegetables. You can add petals from various flowers if you desire, and you can ferment blossoms in sugar water. The taste or aroma will be picked up by the water and transferred to the wine when the water is added to the vegetable wine.

You will, in all probability, decide that many vegetable wines are not for you, but you can also find some unique, and at times highly enjoyable, taste experiences in vegetable wine-making.

If you have a flourishing summer garden, orchard, or vineyard, you can gather enough fruits and vegetables to make a wide range of wines, and all you will have to pay is the cost of sugar.

In fact, you can keep your own bees and use the honey for your cooking and wine-making needs. By nearly all accounts, honey is much better for you than granulated sugar, and it can be very cheap and easy to obtain if you wish to start a beekeeping venture.

17

Using Nuts, Grains, and Honey

Three of the most totally enjoyable and nutritious foods available to us are nuts, grains, and honey. These three fine foods will keep all year with no effort except to protect them from dampness and insects. You don't have to can them, store them in a root cellar, freeze them, dehydrate them, or worry about spoilage under reasonable and sensible conditions.

You can freeze, can, and dehydrate nuts if you wish. Sometimes, in truth, it may be better to do so, because many creatures other than man agree that this particular food is indeed one of life's great pleasures.

If you live in an area where squirrels abound, and if you keep the nuts in an open area, you can almost be assured that sooner or later the squirrels will find a way to confiscate your nuts. Mice are attracted to grains of all types. With a little planning and attention, however, you don't have to worry about invasions.

When you harvest your grains, first clean off all the chaff and then allow the grain to dry either in the sun or in a dehydrator (Fig. 17-1). Very little drying is needed since the grains usually remain in the field until they are nearly ready to use.

You may not feel that it is worth your time and effort to try to grow your own grains unless you have access to harvesting equipment. Most people who own such equipment will not be too interested in hauling their machinery out to your place to harvest 1/2 acre of wheat or less. You may feel that it is too time-consuming to try to harvest it by hand.

Look, though, at two possibilities. If someone near you is growing wheat, for instance, when the harvesting machinery starts to work, go out, introduce yourself, and ask if you could buy a bucketful of the harvested wheat. You will probably find that the harvester, looking at the mountain of harvested grain cleaned and ready for the mill, will not even think of denying your request. He will also be far more inclined to give you the grain than to sell it to you, as long as your requests are modest.

The other possibility is to sow a small patch of wheat, rye, barley, or other grain. When it is

Fig. 17-1. Grain that is cleaned of chaff and other foreign matter is excellent for many types of cooking uses. The grain requires no special storage except to keep it away from mice and insects, as well as moisture.

ready for harvesting, you can simply walk along and strip the grain by the handful and drop it into a bucket. In a few minutes you will have enough wheat to make bread for weeks, possibly months.

GRINDING THE GRAIN

After you rub the grain between your palms, and let the wind blow away the chaff, you can start to grind it. You can also take it to a mill to have it ground, or you can store it in a convenient place and grind it only as you need it.

You can buy a small grain mill that is hand-operated for a fairly reasonable price, and the outfit works well enough that you can grind your own meal and flour easily as long as you can turn a crank and as long as you have the patience to watch the product of your work take its form slowly.

The grinder can be bought for around $100. While we realize that you can buy a great deal of flour for the same amount of money, we have found our mill to be very useful, money-saving, and enjoyable. It takes us about 15 minutes to grind enough flour for 2 weeks' supply, and it takes less than 5 minutes for us to grind the corn we feed to our young chickens each day.

You can grind your wheat or corn while you wait for the cake to finish baking, for your favorite television show to come on, or to pick up your child at school.

You can store the grain before you mill it in any container that will keep it from getting damp or infested by insects. When you grind the flour, grind only about 2 to 5 pounds, and then freeze the flour for 2 or 3 days to be certain there will be no problems with insects.

Do the same with corn meal. As long as no appreciable moisture gets into the ground meal, it will keep for weeks with no preservatives.

Here, in a strong sense, is one of the best arguments for grinding your own bread grains and meals. The federal government allows a certain percentage of commercially prepared foods to be of "foreign nature," as one critic described it. These foreign elements consist of artificial preservatives for the most part, but there are also insect wings or legs, or at times whole insects ground up, rat hair, even rodent manure in foods that pass government inspection. It has been reported that frozen pizza has the highest amounts of foreign matter, but meals and flours are high, especially in unsupervised areas of preparation.

If you grind your own, you can pick out anything that, in your mind, does not belong in the meal. You can remove dark kernels or grains, bits of hull or chaff that might be missed in a commercial operation, and particles of weeds and stems that the commercial preparations might include.

NUTLESS NUT BREAD OR CORNLESS CORN BREAD

If you are concerned that your family might be eating too much flour or corn meal, there is a way to have a bread that is not made of grain but is part of the home-ground meal process.

The bread is delicious and filled with proteins that ordinary breads do not have. It tastes almost exactly like a nut bread but there are none of the excessive calories and cholesterol amounts found usually in nuts.

The Recipe

The bread is made of soybean flour or meal. Start with a small hopperful of soybeans that have been cleaned and then washed thoroughly. After washing the beans, spread them on a cloth and blot them free of all noticeable moisture. Then spread them on a dry surface and allow them to air dry.

You can do a lot of the beans this way at one time, if you prefer. If you prepare several pounds, you can later seal the beans in an air-tight container or hang them in a clean, closely knit bag.

After your beans are ready to use, put them in the home meal grinder and set the burrs on a fine grind. After the beans are ground, put them through a sieve or sifter to get all of the larger pieces out. Grind the larger pieces a second or third time until the beans are finely ground.

Add 1 teaspoon of baking powder, 1/2 teaspoon of soda, 1 egg, 1 tablespoon of flour, 1 teaspoon of salt—or you can omit the salt if you are watching your salt intake—1 cup of buttermilk—or regular milk with a tablespoon of vinegar added to it—and 2 full cups of soybean flour or meal. Mix all of the ingredients well and pour into a black cast-iron pan that has been liberally greased with shortening. Bake for 20 minutes at 400 degrees.

The finished bread comes out in a beautiful cake, and it has a delicious taste, which is richer than corn bread.

The tiny pieces of ground beans in the batter are softened during baking, and they become crunchy like nuts and even taste a great deal like nuts. The nutless nut bread slices well, has a nice, crispy crust, and is good to eat with any meal that you would include corn bread with.

Other Uses for Soybeans

Soybeans, in addition to their use in breads, can be roasted like peanuts, boiled like peanuts or other nuts or beans, made into milk, curdled into cottage or block cheese, and used as a substitute for meat in a wide variety of dishes.

You can dry-roast soybeans the way you would peanuts. Wash and rinse shelled or hulled soybeans, then soak for 5 hours in cold water. Drain, blot dry, and then roast at 225 degrees for about 2 hours. Remove from the oven and eat as they are, or salted, if you prefer.

You can also try various other methods of preparing soybeans to use in soups, casseroles, meat loafs, etc. In fact, a soybean "meat" loaf is delicious, costs only pennies instead of dollars, and is easy to prepare. It is filled with superb food value and very healthful. To make it, just use ground soybeans instead of ground beef. Use eggs and cracker crumbs to bind the ground beans together, along with the other ingredients in the usual meat loaf.

You can also make hamburgers by grinding soy beans and using crumbs, onion chopped finely, and add enough moisture—dry milk reconstituted or even 1/5 of the normal amount of meat—and you can fry or charcoal the burgers as you would all-meat burgers.

PRESERVING NUTS

You can keep nuts in the shells for many months, as long as they don't get wet, but at times it is desirable to preserve nut meats in other ways.

To prepare nut meats, your most important, and often most difficult, job is to find a quick or easy way to get the nut meats from the shells.

You very likely have seen the nutcracker devices often advertised on television. These devices work well but are also expensive, but not in terms of the amount of time you will save and the amount of nut meats you will get.

If you prefer working in the old-fashioned way, you can purchase a nutcracker set or equipment for only a few dollars, or you can use other methods. For instance, you can crack black walnuts or hickory nuts with a hammer and a block of wood. You must first remove the outer hull or sheath, and then simply and carefully hold the nut with your fingers and tap it with a hammer until the shell cracks. Nut picks can be used, but the work is slow, and you crumble or break many of the nut meats.

You can, however, break the shells so that the nut will separate into 3 or 4 pieces, then put the nuts into a warm oven, in front of the fire, or under the wood stove if there is a fire in it. Leave them for 1 or 2 hours. The nut meats will shrink as they dry out and you can remove whole sections with very little difficulty.

If you have a good supply of pecans, put them into a large pot and pour boiling water over the nuts. Let the nuts stand in the water for an hour. The heat from the water will not cook the nuts or damage the taste in any way but it will soften the shells so that you can strip away a shell easily and extract the whole nut meat, or at least half of it.

You can process pecans in the pressure canner or in the water-bath canner if you shell the nuts and then dehydrate or dry them in a 250-degree oven for just long enough to dry out the meats until they are easily broken. Do not cook the nuts. Remove them from the heat if they start to darken significantly. Pack the nuts into jars and close. Process in the pressure canner at 5 pounds of pressure for 15 minutes or in the water bath for 1/2 hour.

You can freeze nut meats by shelling the nuts and placing the meats in an air-tight freezer bag and storing without pretreatment in the freezer.

You can dehydrate or dry nut meats, including peanuts, by shelling the nuts and then spreading the meats or kernels on a drying rack. Leave them until the meats are crunchy and devoid of the green taste. Pack the nuts in an air-tight container and keep on the pantry shelf for months, if you desire.

OBTAINING AND KEEPING HONEY

One of the best of all possible natural foods is honey. It can be very difficult or expensive to obtain, or it can be a very easy and simple matter. If you buy honey in the supermarket, you will pay about $3.50, perhaps more, for a quart of strained honey. If you know someone who keeps bees, you can often buy a quart for $2.00 per quart or $2.00 for one frame of honey that is just as the bees made it.

You can take the frame, put it into a large plastic pan, and a run knife blade around the wooden border. The huge rectangle of honey will fall out in 1 piece. You can cut the honey sections into smaller pieces. Fit them into wide-mouthed canning jars and put lids on the jars. The honey in the comb will seep out and later you will have a jar or several jars nearly filled with honey and comb. You can also cut the comb into small sections and squeeze it by hand as though you were squeezing a sponge, and the honey will drip into the container. You can later ladle it into the jars.

Another way to get your honey is to keep your own bees. If you are allergic to bee stings, do not attempt to keep bees. If stings do not bother you particularly, you can get set up in beekeeping fairly simply.

Setting Up Beekeeping

You can buy or build a bee box or hive with a minimum of trouble. Locate someone who has one and will let you examine the dimensions or examine one in a store and see how it is built. The hive will cost you about $65.00 in the store but you can make one from scrap lumber for only pennies. You can then either buy or make the frames that fit inside, and you can buy the necessary beeswax for a minimal sum.

There are several small booklets that tell you how to take care of bees, but the most important aspect is to protect yourself by being covered by a bee bonnet and several thicknesses of clothing before you enter the hive.

Once you get your first swarm of bees, other swarms will sometimes come to your property and you can box these too. It is good to have the bees

Fig. 17-2. When handling bees or working in the hive, be sure to wear plenty of clothing so that there is no exposed skin for the bees to attack.

Fig. 17-3. You can keep honey in virtually any type of wide-mouthed jar. The honey shown here is one robbing of one hive, and only two supers of honey were taken from the hive—less than half the amount that could have been robbed.

around for cross-pollination of crops as well as for the honey.

Robbing Bees

Plan to rob bees in mid-June or perhaps later, after the first honey flow has been capped. Put on gloves, bonnet, and jacket. Be sure that your trousers are tied to your boots and that all other access ways to your flesh are closed (Fig. 17-2).

Use a smoker to send a few puffs of smoke into the hive opening, and then wait for 5 or 6 minutes while the smoke takes effect. Pry the top of the hive off gently and use a long-bladed knife to cut each frame out and place it into a large pan that can be covered so that the frame will not be literally covered itself with bees.

If you are concerned that you may be taking honey that may contain young bees, you can add a second story to the hive. This second story is called a **super.** It has its own frames but is separated from the basic hive by a **queen excluder**—a wire framework that has openings large enough for worker bees to pass through but which is too small for the queen to get through and lay eggs in the new honey.

Once you have smoked the hive, lifted off the top, and cut out the frames, take them to an enclosed area. Cut the honey from the frames, then pack the honey into jars or cut the combs apart and squeeze the pure honey into the jars.

Once the jars are full, you do not need to process the honey at all; it will keep in a jar that has a tight-fitting lid for months (Fig. 17-3). We have kept honey at room temperature and in unsealed jars for over a year and we have not noticed any dimunition in flavor or consistency.

If you plan to eat the honey straight from the jar (with hot biscuits and butter as your only other food), you may leave the comb in the jar. If you plan to cook with the honey, you may prefer keeping only strained honey.

18

New Uses For Old Foods

It has been pointed out in this book that you can use the peelings, cores, seeds, husks, hulls, shells, rind, and pits of many if not all fruits and vegetables. In your efforts you often find new uses for old foods.

SOYBEANS

Earlier in this book several passages were devoted to one of the oldest foods known to man—the soybean. In addition to the materials presented, however, there are numerous other uses for this very versatile bean.

Similarly, there are many ways in which this food can be preserved, either by itself or with other foods. In an age of weight consciousness and energy attention, this particular food deserves a great deal of attention, not only for its high nutritional value but for its healthful attributes in other respects.

Unlike many other foods, which can be eaten only at particular points of ripeness, the soybean can be eaten as sprouts—only 4 or 5 days old—as young green or unripe beans, and when fully matured or bush-ripened.

While other beans have been a staple of the American diet, however, the soybean has been noticeably absent until recent years. Today, it ranks as the top row crop in the United States, and all indications are that by the late eighties it will gain a place of great importance in American diets.

Why? First, it grows well in many areas of the country and does not require highly specialized care. Second, it produces an abundance of beans as a garden crop. Third, it has probably as many uses as any other vegetable.

The list of reasons could continue indefinitely. The soybean has a high fat content and contains calcium, phosphorus, iron, copper, magnesium, and potassium, as well as other nutrients.

Fat and protein content varies between types of soybean, but generally the protein content is about 35 percent, and the fat content runs slightly under 20 percent.

If you choose to grow your own soybeans, there are many so-called garden varieties to choose from. These varieties are usually greener than the commercial soybean.

The commercial soybean can also be grown in the backyard garden and eaten safely and enjoyably; it is highly doubtful, in fact, whether many people could tell the slightest difference between garden and commercial varieties.

If you prefer not to grow your own, you can buy a sack or bushel of soybeans directly from farmers, harvesters, health food stores, and some supermarkets.

Buying from the health food store is perhaps the most expensive method of obtaining soybeans. The least expensive way is to approach the harvester as he fills the huge truck bed with the beans and offer to buy a bushel, which he will likely sell to you at a very low rate. Supermarket prices will also be considerably higher than those of the farmers' supply store. If, however, you buy at a farmers' supply store, be sure to ask if the beans have been treated with chemicals for planting purposes. If so, do not use the beans for food; if not, they are perfectly fine for table use.

If you grow your own, be prepared for meticulous work, since the beans grow low to the ground and are uncomfortable to pick. Once picked, they are tedious to shell, since the beans and the pods are smaller than many other bean varieties. When buying the beans, buy those already shelled and ready to prepare. It is worth the small amount of extra money to save the time and work needed to prepare the beans for canning or drying.

Unfortunately, if you buy the harvested beans, you will be getting only the bush-ripened ones, and you may prefer to try working with younger, more tender beans.

Canning Soybeans

Your first step in canning soybeans is to shell the beans, which can be done rather easily by soaking them in water that has been heated to a rolling boil. After you drop the beans into the water, time the soaking fairly closely and limit the time to 5 or 6 minutes.

Pour off the hot water at the end of the soaking period. Allow the beans to cool until they can be handled with comfort and simply tear open the shell and squeeze the beans out. Be sure to select young, tender, but firm beans for canning.

After shelling the beans, drop them into a pot of boiling water and cook for 4 minutes, and then pack them loosely into jars. Pint jars are probably better than quart containers for canning soybeans since a pint of these nutritious beans will feed several people.

After filling the containers, add 1 1/2 teaspoons of sugar to each pint of beans—sugar helps the beans to retain their original flavor—and, optionally, 1 teaspoon of salt for each pint.

Pour boiling water over the beans and close the jars with preheated lids and rings. Tighten firmly. Process the jars of beans at 10 pounds of pressure for no less than 30 minutes per pint. Do not attempt to can soybeans in a water-bath canner, since it is a low-acid food.

When you open the beans to serve, be sure to boil them for 15 to 20 minutes before eating. Do not even taste the beans, or any of the low-acid canned foods, until after the boiling period after opening the jars.

Freezing Soybeans

Soybeans can be frozen quickly and easily. Start with young, tender beans that are as fresh as possible. As with other vegetables, it is best to process immediately after picking if possible.

Put the beans, still in their pods, into a vessel containing boiling water and blanch for 5 to 7 minutes. Do not overcook. Remove the beans from the water and allow them to drain and cool for several minutes. Hull the beans and drop them into the freezer bags or cartons. You do not need to add sugar or salt.

To prepare frozen soybeans for table use, drop the beans, still frozen, into a vessel of boiling water and cook for 10 to 12 minutes. Drain, add seasoning, and serve as a separate dish or mixed with other vegetables.

Do not thaw the beans prior to cooling and they will retain their crisp, nutty flavor. You may prefer to cook the beans in unsalted water. You may add butter or margarine after the beans have been removed from the heat if you wish.

Drying Soybeans

Drying soybeans is about as simple as any part of home food preservation can be, and we recommend drying beans, especially soybeans, since the dried food can be used in so many ways.

You may allow the pods to remain on the bushes or plants and dry by sunlight and air. In fact, you may leave the pods on the plants until well into the winter. In the South many commercial bean growers do not harvest their crop until after New Year's Day, although this practice is somewhat injudicious since the rainy winter weather may cause sprouting or decay.

Still, you may let the pods dry on the plant and leave them safely until December 1, or you may pick and dry as your time permits and set your timetable so that you will have the crop gathered, if it is a large one, by early winter.

If the beans somehow are not totally dried, you may spread them, still inside the pods, in a sunny but sheltered area to complete the drying process. You may also hull the beans and spread them on a protected surface to air dry. When you store the beans, take care not to allow moisture to get into the container, or the beans may start to sprout, especially in the bottom of the bucket or sack where it is dark and moist.

Roasting Soybeans

Once you are ready to use the beans you have dried, canned, or frozen, you will find that they are delicious when prepared in a wide variety of methods. The frozen or canned beans are best when prepared like any other beans and can be used in casseroles, soups, salads, and various other vegetable dishes.

It is the dried soybean, however, that is almost a miracle of versatility. The number of possible uses is astonishing and rewarding.

For instance, you may choose to use roasted soybeans instead of peanuts as snacks or ingredients in various main dishes or desserts.

Start by washing the beans in cold water, then place the beans in a pot or other vessel and cover them with a generous amount of water, enough so that the beans will not absorb all of it during the soaking period. You may simply pour the water over the beans, cover the vessel, and let the beans soak for 8 to 10 hours or overnight. You can also pour boiling water over the beans and allow them to soak for 4 or 5 hours (Fig. 18-1).

After the soaking is completed, drain the water and spread the beans on a cookie sheet or similar surface and allow them to dry for an hour or so. When the beans are dried, you can roast in one of two ways: deep fat preparation or oven roasting (Fig. 18-2).

You can roast the beans exactly as you would prepare french fried potatoes by heating cooking oil to the boiling point, adding the beans and allowing them to cook until they are a golden brown or a shade lighter—usually the cooking time is about 10 minutes.

Remove the beans from the heat and drain off the oil. Let the beans air dry for several minutes and then, if you prefer salting, drop the beans into a plastic carton or paper bag, shake salt into the container, and shake the beans vigorously until the salt has been evenly distributed.

To oven roast, soak, drain, and air dry. Place the beans in a shallow pan or on a cookie sheet and roast in an oven at 400 degrees for 20 minutes, or until the beans are a golden brown. You may salt, if you prefer, or leave natural. Store the roasted beans in a closed container unless they are to be eaten within a day or two.

If you prefer, you may use the roasted beans in various recipes for pecan pies or nut breads and cakes. Simply follow the original recipe but instead of using nuts substitute soybeans that are crisply roasted. You can also mix nuts and beans.

Soybean Candy

For a delightful treat, soak soybeans by pouring boiling water over them and allowing them to sit for several hours. Then use 2 cups of beans and you

Fig. 18-1. Three cups of soybeans, after soaking, expand to fill a large space in a huge container.

can make a delightful candy that is almost identical to peanut brittle.

Put the soaked beans on a cookie sheet after they have been well-drained, and roast the beans for 15 minutes at 425 degrees. Stir the beans occasionally while they are roasting. Remove the beans for the oven, place them in a pot, and add 1 cup of granulated sugar, 1/2 cup of light corn syrup, and a pinch of salt. Cook for about 8 minutes or until a light brown color. Then add 3/4 teaspoon of baking soda and 1 1/2 tablespoons of margarine. Stir well, and while the mixture is foaming, pour on a greased cookie sheet to cool.

You may also melt chocolate and stir in roasted beans for a chocolate-covered peanut substitute that is less fattening.

Soy Milk

Some people are allergic to cow's milk or are reluctant to ingest the fats present in milk; nevertheless, these people—and others who dislike paying the price of milk—still need milk for many types of food preparations.

Again the soybean provides an answer to the problem. Soy milk is easily made and stored. To make the milk, use 1 pound or 3 cups of dried soybeans that have been hulled. Wash the beans and pour boiling water over them. Allow the beans to soak overnight or at least 4 or 5 hours. If you do not use boiling water you may soak the beans for 8 to 10 hours at room temperature and achieve the same results. Be sure to cover the beans generously so that when the beans swell and expand with water absorption there will be adequate water to keep the beans covered.

After the soaking has been completed, drain the water off the beans and then cover the beans with fresh water and boil for 15 minutes. Drain a second time and allow the beans to air dry by spreading them in a shallow pan and placing the pan in direct sunlight or in a warm oven.

When the beans are fairly dry, use a blender or a good grinder or chopper to grind the beans as finely as you can. Place the beans then into a cloth that is porous and tie the cloth so that the beans cannot escape.

Put the bean-filled cloth into a large pan or pot and then pour half a gallon of warm water into the vessel. Use both hands to knead and press the bean pulp until the "milk" begins to soak from the pulp.

Continue kneading for 10 minutes, then squeeze the pulp bag firmly to remove any remaining milk. You will have 1/2 gallon of milk at this point. Your final step is to bring the milk to a rolling boil and hold the temperature at boiling for 15 minutes. Cool the milk and pour it into appropriate container and store it in the refrigerator.

To use the milk, substitute soy milk in equal measures for regular dairy milk. You can use it for baking custard, pies, breads, and various vegetable dishes, and you can use it in preparation of hot chocolate, soups, gravies, and sauces.

For instance, if you wish to bake biscuits, use 2 cups of self-rising flour, 1/4 cup of shortening, and enough soy milk to make the dough moist and slightly tacky. Roll the dough, after kneading, in

Fig. 18-2. You can roast soybeans in the oven just as you would roast peanuts, and the results are very similar.

enough flour to coat the dough so that it can be handled easily, and then cut biscuits from the dough. bake at 425 degrees for 10 minutes. The bread will brown a slightly deeper color than milk biscuits will, and the consistency of the bread will be very rich. These biscuits, incidentally, can be made several days ahead of time and warmed in a 300-degree oven, and they will still taste very fresh. They will be extremely filling since you get protein as carbohydrates.

If you prefer to use plain flour, add 3 teaspoons of baking powder and 1/2 teaspoon of salt to the flour before adding shortening and soy milk.

There are many other recipes in which soy milk may be used readily.

Bean Pulp Dishes

With the bean pulp that was left over from the soy milk you can make delicious ''meat'' loaf.

You will need 2 cups of the ground bean pulp, 4 slices of bread toasted and crumbled, 1 egg, 1/4 cup of chopped onion, 4 tablespoons of melted margarine, 1/3 cup of catsup, and 1 beef bouillon cube. The bouillon and the onion are optional. Combine all ingredients, then add 2/3 cup of plain yogurt or cottage cheese. Place the ingredients in a baking pan and bake, covered, for 40 minutes, then uncovered for 20 minutes. Serve as you would serve a meat loaf.

The dried pulp can also be used in pies or cakes that require nuts, in salads, and in a wide variety of soups, sandwiches, or wherever hamburger would be used. The pulp is excellent when fried in patties for sandwiches, in casseroles, in hash dishes, and in sauces.

If you wish, you can dry the pulp thoroughly and grind it even finer with a hand-operated mill. Use the ground pulp as you would use corn meal or flour. If you don't dry the pulp well it will have a tendency to clog the burrs and other mechanisms of the grinder. If you prefer, too, you may mix ground soy pulp, with flour or meal, usually at a rate of 3 tablespoons to 1 cup of flour or meal. You may also use 1 part soy pulp to 2 parts flour or meal.

You may also freeze the pulp and keep it for later use. It should be placed in a freezer carton and tightly closed before you place it in the freezer, where it will keep for months in the freezer. Similarly, any dishes you make from the pulp can be frozen to be served later.

Soy Sprouts

For several years bean sprouts of all types have been very popular items for the salad bar or as vegetable dishes, and soy beans can be used with excellent results for sprouts. Use fresh or current crop beans—not over a year old for best results—that are well-dried.

Select only whole beans—those that are split or cracked will not likely sprout—wash beans thoroughly; 1 cup of beans, perhaps 1 and 1/3 cup at most per quart container. Soak overnight or 8 hours or so. Add 3 cups of water for each cup of beans so that there will be no danger that the beans will soak up all the water.

After the soaking has been completed, place the beans in a quart jar or similar container and cover with a cheesecloth or similar fabric. You may simply drape a thin cloth over the mouth of the jar. Then store the jar in a dark place.

For best results, lean or tilt the jar slightly. Once every 3 hours or so take the jar of beans from the cabinet and rinse the beans well. Be sure to do this at least three or four times each day until the sprouting starts. Frequent rinsing will help to keep mold from forming on the beans. Do not wash vigorously enough to damage the softening beans, however.

Wait until the sprouts are about 3 inches long, or slightly shorter, before using the sprouts. If you wish to use sprouts in salads, you will probably prefer to boil the sprouts for 4 or 5 minutes, then refrigerate to keep the sprouts crisp. The sprouts may be used in various casseroles, Chinese dishes, or simply cooked along with beans as a vegetable dish.

The growth time required, after the beans first begin to sprout, is 5 or 6 days. To ensure a ready supply of sprouts at all times, start a new jar about every 2 or 3 days. If you do not plan to eat sprouts often, you may prefer to grow only 1 jar at a time. Be sure to refrigerate the sprouts as soon as they are ready to eat, but do not freeze them. It is so much easier to keep the dried soybeans for sprouting than it is to prepare the beans for freezing.

Soy Oil

If you have an abundant supply of beans, you may wish to try pressing soy oil from the beans. Soak beans for 8 to 10 hours, then cook slowly until the beans are soft and tender. They will not become mushy as great northern beans or pintos will, however. When the beans are ready, drain the water from the beans and place the beans in a grape or fruit press that has been washed thoroughly.

Slowly apply pressure so that the oil can be pressed out gently and caught in clean containers. If you do not have access to a press, you can spread the beans in a flat pan and use a rolling pin to crush the oil from the beans. Drain the oil into containers.

According to the American Soybean Association Bulletin, soybeans provide the most abundant and least expensive sources of food or cooking oil. Three-fourths of all margarines, 2/3 of salad and cooking oil, and nearly 1/2 the cooking fats are made from soy oil. The association reports that the oil is high in polyunsaturates, so it is ideal for use in salad dressing, mayonnaise, shortening, margarines, and cooking oils. The oil can be used for any purpose that commercially prepared cooking oils would serve.

One of the most attractive features of soybeans is that the beans themselves require no special treatment for preservation. After the beans dry on the plant, all you need to do is keep them in a bucket, sack, or any other container that will protect the beans from moisture and unintentional sprouting. The beans keep well in warm weather, cold weather, and in sunshine or shade, in light or dark storage areas.

Tofu or Soybean Cottage Cheese

For thousands of years Oriental people have enjoyed a food called ***tofu*** (tow-foo), which is a type of cottage cheese made from soybeans. To make tofu, soak fresh beans for 8 hours, then cook slowly until they are rather soft and mushy. Allow the beans to stand until the curdling starts. (See the chapter on cottage cheese-making for details on how to produce curdling.) Separate the curd from the whey. Keep the curds, or curdled soy milk, and press it into molds or shape it into squares or cubes. The resulting hardened curds, referred to as tofu, will have numerous uses in food preparation.

First of all, tofu is virtually tasteless. It will have a faint bean odor and taste, but the fact that it is bland can be an advantage, since so many other foods can be mixed with tofu without having it interfere with the overall taste of the other ingredients.

Since there will be a great deal of water added to cook the beans in, 20 pounds of soybeans will produce 50 pounds of tofu. The tofu is 95 percent digestible and, like many other soybean products, is cholesterol free, low in calories and very high in vitamin B1 and iron, but very low in sodium.

When the tofu is ready and has been shaped in blocks, you can wrap it in air-tight bags and freeze it until you are ready to use it. It will keep for several days in the refrigerator, if you prefer.

Tofu Dressing. To prepare tofu for meals, you may mix various spices, relishes, pickles, peppers, and salad dressings. A typical dressing might consist of 8 ounces of tofu, 2 tablespoons of mayonnaise, 1/2 tablespoon of prepared mustard, a sprinkle of garlic powder, and salt and black pepper to taste. Parboil the tofu for 5 minutes, then add all the other ingredients, and serve on wheat toast or similar bread.

Tofu Frosting. You may wish to make a cake frosting. To do so you will need a pound of tofu, 1 stick of margarine, 2 teaspoons of vanilla, and 1/2 cup of light, strained honey. Cook together all ingredients in a double boiler until the mixture thickens, then blend or use beaters to mix until it is creamy and smooth. It can then be used, or it can be refrigerated until you are ready to use it.

Fried Tofu. You may also wish to try sliced fried tofu. Melt 2 pats of margarine in a frying pan and add a tablespoon of Worcestershire sauce. While the pan is heating, slice tofu 1/2 inch thick and roll in flour, then fry the tofu in the margarine sauce mixture until it is the color of light toast. It is ready to eat when it has been browned on both sides.

Soy Milk Ice Cream

Ice cream is easy to make, and the results are palatable, but the soybean taste is still perhaps too apparent for many people's tastes. You can taste the soybeans in biscuits made from the milk, but here the taste adds to the enjoyment of the food.

All you need is 3 cups soymilk, 1/2 cup of honey or a cup of sugar, 1/3 cup of cooking or salad oil, and a pinch of salt. You can blend all of these ingredients and then freeze, and the result is 1 quart of vanilla ice cream, minus the vanilla flavoring. If you don't enjoy the taste of vanilla, you might want to add only 1/2 teaspoon of vanilla flavoring. If, on the other hand, you enjoy the taste exceedingly, add 1 1/2 tablespoons to the ingredients. One tablespoon of vanilla makes a distinctively tasting mixture but the flavor is not overpowering.

We have found that mixing soymilk and regular milk makes a much better, for our tastes, ice cream. You may wish to try mixing 1/3 whole milk, 1/3 dry milk, and 1/3 soymilk; or you might prefer 1/3 cream, 1/3 whole milk, and 1/3 soymilk.

It becomes expensive in time and ingredients if you make a gallon of ice cream and then don't like it well enough to eat it. It takes an hour, easily, and probably more to make a freezer of ice cream.

We use a basic recipe for all flavors of ice cream, and we simply add an extra ingredient if we desire. Our basic recipe calls for 4 cups of milk, 2 cups of sugar, 1/2 teaspoon of salt, and 4 cups of light cream, plus 2 tablespoons vanilla flavoring. Scald the milk, add the sugar and salt to the milk, and stir until they are mixed completely. Add the cream, stir in the vanilla, and place the freezer tank in the refrigerator to cool the mixture before you start to crank.

You may wish to try substituting the soymilk for the whole milk or the cream, if you want a low-fat ice cream—or ice milk, more properly. If you choose, you can peel, slice, and mash 4 ripe peaches to add to the mixture, or you can add 2 cups of strawberries or whatever other fruit you especially enjoy. You will be adding to the volume of the mix by using extra fruit, so you may need to reduce the amount of milk accordingly so the freezer will not run over when the freezing mixture expands.

FLOUR FROM SWEET POTATOES

While virtually everyone is familiar with the basic flours made from wheat, rye, and other grains, few people seem to realize that flour can also be made from several vegetables normally not associated with flour. Generally, these vegetable flours are not typically used for gravies, cakes, or pie crusts; instead they are used for interesting variations in breads.

You may not want to use these flour types for biscuits to be eaten with preserves, jams, and jellies, since some of the breads resulting from unusual flours tend to be sweet already and need no jellies, while others are more vegetable-tasting and tend to clash with sweet breakfast spreads.

You can slice sweet potatoes into thin ovals and dehydrate the slices until they are crisp. Then break, cut, or chop the slices into small pieces and feed them through the food grinder on a fine setting. The resulting "flour" can be used for several types of breads.

Soybean Bread

The simplest bread can be made by using a basic mixture of wheat flour and ground potato flour. The proportions used can be varied to meet your own tastes and needs, but a normal amount would be 1/3 sweet potato flour to 2/3 wheat flour. You can make a basic bread by following your favorite recipes, and you can add chopped nuts and a variety of spices to the ingredients.

A basic recipe uses 3 tablespoons of yeast, 1/2 cup of honey, maple syrup, or pancake syrup (you may increase the amount by 1/3 cup if you wish a sweeter bread), 3 cups of milk, 1 cup of chopped raisins, 1/2 cup of chopped nuts or soybeans, 7 1/4 cup of wheat flour, 1/2 to 2/3 cup of sweet potato flour, and 4 tablespoons of cooking oil. You may add 1/2 teaspoon of nutmeg, cinnamon, or other spices, if you like.

Mix all dry ingredients and stir until they are

uniformly mixed, then add the milk and other ingredients slowly, mixing well. Knead the dough well and lightly brush the shaped loaves with soy oil, melted margarine, or milk. Allow the loaves to rise 40 minutes and bake at 350 degrees for 50 to 60 minutes.

Soybean Pancakes

You can also make a basic pancake flour by mixing equal portions of wheat flour and sweet potato flour. Use self-rising flour and omit baking powder; if you prefer plain flour add 1 heaping teaspoon of baking powder. You can stir into the batter other ingredients such as chopped nuts or blueberries or other fruit, then prepare as you would prepare basic pancakes.

If you use fresh fruit, you may want to make the pancake batter slightly thicker than you would normally, since the cooking fruit exudes juices that will thin the pancakes to a slight degree. Fresh fruit in the batter will cause a thinning even before cooking.

NEW USES FOR PRESERVED POTATOES

In the late summer when you typically store Irish or white potatoes, you normally think about french fries and baked or mashed potatoes in the winter. There are several other uses for Irish potatoes, however, whether dehydrated or stored in a root cellar.

If you plan to use root-cellar potatoes, all you need to do is peel, wash and dice the potatoes. If you intend to use dehydrated potatoes, dump them into a cooking vessel and add enough water to cover the potatoes. Wait for an hour or so to allow the potatoes to absorb the moisture that was removed during dehydration. If they absorb all of the water quickly, add more. If there is a great deal left over after the potatoes have plumped, you can always drain it.

Once the dehydrated potatoes are ready, cook them slowly until they are tender enough to mash easily. Use a fork, beaters, or a blender to beat the potatoes into a smooth texture. You are now ready to use the potatoes in a new recipe.

Potato Bread

You can make delicious potato bread or ice box rolls from the following recipe. You will need 8 cups of plain, not self-rising flour, 2 teaspoons of salt, a package of baker's yeast, a cup of warm water, 1/2 cup of sugar, 2 eggs, slightly more than 1/2 cup of milk that is moderately warm, and a generous cup of the potatoes you have prepared, plus 4 tablespoons of shortening.

Start by dissolving the yeast in a small amount of warm water. The water should be slightly cool to your hand, thus slightly below body temperature. Allow 10 to 15 minutes for the yeast to dissolve completely.

While the yeast is dissolving, mix the sugar, shortening, eggs, and potatoes together and blend until the ingredients are smooth. Slowly add the flour and work it into the mixture very thoroughly. Add milk at this point, but do so slowly to be certain that you are not adding too much. Sometimes the size of eggs can make a recipe batter thinner than smaller eggs would have, as can the softness of margarine as opposed to stiff margarine or cold shortening.

When the dough is damp but not sticky you can start to knead it gently. Continue to knead for 12 minutes, perhaps slightly more, and then allow the dough to rise until it has doubled its volume. Knead very gently again before you refrigerate it.

If you wish, you can refrigerate the dough for several hours, but it is probably better to bake the bread within an hour after mixing, since cold potatoes, even when mixed into a dough, have a slight tendency to start to become crisp and coarse. As you make the rolls, place them in a slightly oiled or greased pan. You can instead bake the entire batch of dough as a loaf. Use the same time and temperature as you would for any other loaf bread or rolls.

More Potato Bread

For another Irish potato bread, try using 2 1/4 cups of mashed potatoes, a generous 1/4 cup of brown sugar, 1 1/2 teaspoons of salt—you may decide to use 2 teaspoons of salt if your potatoes are dense rather than thin and watery—1 1/2 cups of lukewarm milk—never add yeast to milk batter that is more than lukewarm—1/3 cup of shortening, 7 or 8 cups of whole wheat flour, 2 eggs, and a package of yeast which has been dissolved in a cup of warm water. Work the shortening into the dough well before continuing.

After the yeast has dissolved, add it to the lukewarm milk and stir gently until the two are well-mixed. Now slowly add the whole wheat flour, a large spoonful at a time, and stir until the dough starts to thicken with the flour volume. Keep stirring until no large lumps are visible, then add the potatoes and stir until the mixture is again well blended. Add the remainder of the flour and stir again until smooth. Mix the sugar, salt, and eggs well into the dough. You can add the sugar, salt, and egg to the milk and yeast, if you prefer, and mix all of the wet ingredients at the same time for ease in working the dough.

When all ingredients have been added and mixed well, knead the dough for 15 minutes. Then place the ball of dough into a greased bowl and cover the dough with a light brushing of melted shortening. Cover the dough and allow it to rise until it has doubled in size. Knead it very gently but do not mash or press the dough. Simply work it with your fingers as if you were giving a very light massage with your fingertips only.

Divide the dough into two sections of equal size and let it rise again until it is double its volume. Bake in a greased pan for 1 hour at 375 degrees.

More Preserved Potato Uses

When you start to dehydrate potatoes, you will find that it is well to slice the potatoes very thinly by using the slicer of a pickle cutter or the one on the side of your metal grater. Any food chopper will work well as long as it will slice the potatoes very thin.

When you are ready for homemade potato chips, all you have to do is heat the cooking oil to a near-boiling point, place the chips into the hot oil, and cook them until they are a light golden brown. Remove the chips from the oil, and place them on paper towels to drain. Salt them to taste and serve immediately.

Potato Cakes

You can make a fine side dish by grating a root-cellar potato or by mashing plumped, dehydrated potatoes. Grate a small onion, if desired, and, if the potatoes are very thin or watery, add enough flour to thicken the potatoes. When you have the texture right, flour both sides of the potato cake that you shape with your hands into a palm-sized oval and fry it in a medium hot pan with melted shortening or margarine in it. Margarine gives the potato cake a golden color but you must stay near the frying cakes in order to keep the margarine from overheating and burning. Again, do not fry in a hot pan. The cakes can be refrigerated and reheated for leftover uses or they can be frozen and heated later.

Whole Potato French Fries

When you are digging your Irish or white potatoes you will have numerous very small potatoes that did not have a chance to develop before the plant died. Do not discard these tiny potatoes. No matter what size, the potatoes can be delightfully utilized in the kitchen.

There is much disagreement concerning the best times to dig potatoes. It is argued that potatoes should be dug as soon as the peak growth period has passed. Proponents of this argument insist that rains and summer heat will cause an unusually large number of the potatoes to decay prematurely.

Others point out that potatoes will keep in the soil as well as in the root cellar, perhaps even better than in the cellar.

Our feeling is that you should start to dig when you need potatoes. We have experienced a great deal of crop loss when we waited too late to dig, however, partially because of decay and partially because of insect damage.

You can dig under a plant and rob larger spuds early in the season. When the plant dies, the potatoes cannot continue to grow; so if you have a good place to keep them, you might as well dig and store them. If your storage area is not better than the soil, you may prefer to leave them in the ground.

If you dig early, almost as soon as the plants die, you will find the small potatoes are firm, hard, and crisp. Collect them into a bag, wash them carefully to rid them of all dirt, and store them in a clean container. When you wish to have french fries, you can simply drop the small potatoes into the cooking oil and in a few minutes you will have delicious fries that are quite possibly superior to those derived from big potatoes. You don't have to worry about scraping or peeling the small ones, since the skins are so tender that they are not noticeable.

MAKING MAYONNAISE

If you have chickens, you can make your own mayonnaise quickly, easily, and inexpensively. You will need only 1 egg for each batch of mayonnaise, however; so if you must buy eggs the expense is still minimal. Many users of homemade mayonnaise insist that it is superior to the commercial varieties in many respects.

The ingredients are simple: 1 egg, 1/2 teaspoon of salt, 1/2 teaspoon of sugar, 1 tablespoon of prepared mustard, and 1 cup of salad oil or soybean oil. The sugar and salt are optional.

There is only one crucial factor in making of mayonnaise: the egg absolutely must be room temperature. If the egg is too cool, it will not cause the mayonnaise to become fluffy, and it will not hold its consistency.

Put the first five ingredients into a blender and run until the mixture is smooth. While the blender is still running, pour in the salad oil. Continue to blend until the mayonnaise is thick and smooth. That's all there is to it.

19

Making Soap

In all activities concerning the growing, preserving, and enjoying of food, there seems to be a necessity for some waste, no matter how trivial. The best gardeners usually are those who have sought and found ways to eliminate more of the waste products.

Little has been said or written about what to do with certain wastes from meat preserving, especially the greases and fats, that apparently have little or no uses except for seasoning certain foods or preparing some gravies. The waste generally goes either to the family pets or to the garbage can.

The grease and fats that are left after the meats have been fried have a very definite use, however, though not necessarily as a food. You can make delightful soaps from the fat, grease, or lard from any kind of meat. You can even use vegetable oils to make soap.

The effort requires little time and virtually no money, and for your troubles you can have huge bath-sized bars of soap for less than 5¢ a bar.

Better still, you can also have soaps that are scented with your favorite perfumes, and you can have a choice of size, shape, and color. Soaps made at home can be left in the bar shape, or they can be grated into soap flakes for use in washing dishes or clothes.

You will need only a few very inexpensive items: lye, water, grease or lard, and, optionally, borax, ammonia, turpentine, and perfume. You can make wonderful soap with only water, grease, lye, and some type of scent or perfume. There are 2 ways to make basic soap from left over fats: with commercial lye and with homemade lye.

MAKING LYE

If you decide to make your own lye, you will be surprised how simple and easy it is. All you will need is a hollow container or a hopper with a drain spout and some ashes and water.

One excellent type of ash holder is a small hollow log. If you can find a dead sourwood tree with a trunk 6 to 10 inches in diameter, you can cut a 5-foot section of the trunk and clean it so that all dead leaves, twigs, and other debris are removed. Once the trunk is cleaned, saw one end so that it is fairly straight and nail a board over the end. You want a fairly close fit but not a perfect one, since you will want water to drip past the board.

Once the end board is in place, stand the log on its end with the board end on the ground. Drop in enough small rocks, pebbles, or gravel to make a drip space of about 4 inches. This part of the hopper allows water to seep through, but the gravel or rocks will stop any large solids.

When the rocks are inside the log, get a handful or two of dried grass, straw, or similar material and shove it into the top opening. You may need a long stick, hoe, or rake handle to help you shove the grass down to the gravel. The purpose of the grass is to keep ashes from washing down with the water.

You are ready to fill the log with ashes at this point. It is best if you have ashes from such hardwoods as oak or hickory. Shovel the ashes into the log until it is filled to within 4 or 5 inches of the top.

Now prepare the log so that it will stand or hang erect or at a sharp slant. There are several satisfactory ways of achieving this position. One way is to tie a strong rope around the top of the log and to suspend the log from a beam, rafter, or tree limb (Fig. 19-1).

You can also cut some small saplings or use old 2-by-4s and construct an X-shaped arrangement. Lay the top end of the log in the crotch of the X and set the bottom end on top of a cement block or small chunk of wood.

Your log hopper should be so positioned that there is room to put a large glass or plastic jar to catch the drippings from the hopper. When all is ready, you can start to pour water into the top of the log. Pour slowly so that the water will not overflow before it can soak into the ashes. When you have poured in about 2 gallons of water, wait until the water has a chance to work its way through the log. It will emerge at the bottom end as a dark brown liquid. This is your lye water, which is instrumental in making soap.

If you pour in 2 gallons of water, you will get back 1 gallon and perhaps 1 or 2 quarts. After you have collected enough water to fill your container, you may wish to pour the lye water back into the log and let it run through 1 or even 2 more times.

Fig. 19-1. Suspend the hollow log over a glass or plastic container to catch the lye water. Inside this cast-iron pot is a gallon container to hold the water, but the water never comes into actual contact with the metal.

You can also empty the ashes and put in fresh ones and run the water through the fresh ashes. Each time it passes through the ashes it becomes stronger and stronger.

Two traditional methods of measuring the strength of lye water are to drop an Irish potato into the water or to lower an egg slowly into the lye water. You want to see how much of the egg or potato will float. If both sink to the bottom, the lye water is not yet strong enough. You have the proper strength when an egg or potato will float high enough that a circle about the size of a quarter remains above the water surface.

You must be sure, however, that your potato is not "green" or that the egg is not spoiled. A rotten egg will float readily even in tap water, but a newly dug potato is too heavy to float even in extremely strong lye water.

Do not under any circumstances allow the lye water to come in contact with your eyes, skin, or clothing, and keep it out of reach of children. Lye that is ingested can be very harmful and even fatal.

If a hollow log is not available, you can make an ash hopper from a few old boards. Start by nailing two boards in a V shape to form a trough. The boards should be fitted well so that water will not run through the joint. You can actually nail the boards at right angles and then turn them so that the joint is at the bottom. The boards should be about 3 to 4 feet long. If the joint is not tight enough you can use a caulk or compound for sealing wood.

Next, nail boards vertically from the sides of the trough. Leave a little space at one end for a drip spout, however. The vertical boards should be about 3 feet high. When this part of the job is done, nail horizontal boards across the front and back so that ashes will not fall through.

Your final task is to fashion some sort of stabilizer so that the hopper will not tip over. You can nail a strip of wood from the top of the ground on each side, but allow enough distance so that the hopper can be positioned on top of cement blocks so that a vessel to capture the lye water can be placed under the spout or drain.

When all is ready, put the hopper in place and place a glass or plastic jug under the drain or spout (Fig. 19-2). Fill the hopper with ashes and either pour in water or allow rainfall to drip through the ashes. Be sure that the container collecting the lye water is covered, though, so that pure rainwater does not fall into the opening.

MAKING SOAP WITH LYE WATER

Once you have collected your lye water and tested it for strength, you are ready to make soap, if you have a supply of grease.

If you do not already do so, it might prove helpful and economical to start to pour the grease and drippings from all cooked meat into a container. An empty shortening can with the plastic lid will make an ideal grease collector.

When you finish cooking, pour the grease through a piece of cloth or sieve so that you will not be collecting bits of meat; the purer the grease, the easier it is to make soap. It does not matter if the grease is mixed from beef, pork, or mutton drippings.

When you are ready to make soap, put the grease—about 2 pounds of grease for each gallon of lye water—and the lye into a cast-iron pot or galvanized container, never an aluminum vessel. Start the mixture heating. Stir slowly while the soap cooks and keep it on the burner until the soap has started to thicken to the consistency of very thick pea soup or, better still, to a paste.

When you have cooked the soap until it is thickened, you are ready to pour it into molds, unless you prefer to scent the soap. If you have very hard water you may wish to add borax so that the soap will lather well.

MAKING SOAP WITH COMMERCIAL LYE

If time or facilities do not permit you to make your own lye water, you can buy a can of commercial lye at the supermarket for about 1 dollar a can. To make soap with prepared lye, you do not need to cook the mixture at all. All you do is pour the contents of the lye can into a plastic bucket and add 5 cups of cold water.

Be very careful. Concentrated lye is dangerous, and you can receive very painful burns unless you exert great care. Lye mixed with cold

Fig. 19-2. Fill the completed hopper with ashes, pour in water, and let the water seep through the ashes. Lye will drip into the container positioned under the spout.

water produces intense heat for several minutes. Temperatures can range as high as 200 degrees almost instantly, and the white fumes that rise can cause coughing and eye irritation.

If you will simply pour the water into the lye slowly and keep your face from over the mixture, you will be in no real danger of any sort.

After the water is in the lye, and you have stirred until the lye is completely dissolved, set the container aside and allow the lye and water to cool. Be sure to stir with a wooden spoon or a wooden strip or stick. Do not use a metal spoon of any sort.

You should at this time melt your lard or grease and measure to see that you have 11 cups. You can buy commercial lard if you don't have a supply of grease. At hog-killing time you can easily locate farmers who have more lard than they can use, and they will sell it very cheaply or even give it away.

After you have melted the lard or grease, you may decide to strain it through a piece of cloth to remove any materials that you do not want to have in soap. Set the lard aside so that it, too, may cool.

A workable idea is to mix lye and water and then melt your grease as soon as you can in the morning. Let it cool while you are eating breakfast or tending to other chores. In cooler weather, of course, the two ingredients will cool very quickly.

When the temperatures are about 75 degrees, you are ready to begin work. If you don't have a thermometer that you wish to put into the lard or lye water, just place the palm of your hand against the outside of the two vessels. If the pails feel warm to the touch, let them cool longer; if they feel only slightly cool to your hand, you can begin mixing the two materials.

If you are in a rush to complete the soapmaking, you can place the two buckets into a larger vessel partially filled with cold water. You may also mix the two at higher temperatures, but the soap will not begin to thicken until the temperatures begin to cool.

When you are ready, use a mixing cup, glass, or other handy, small container and pour the melted grease slowly into the lye mixture. Stir slowly while the grease is being added. You must keep the stirring going at an even rate during the entire process. If you need to stop for a minute or two there will be no problem, but don't leave the mixture for 1/2 hour or even 10 minutes. Keep adding grease until you have put 11 cups into the lye mixture.

As you stir you will notice that the two ingredients begin to fuse into one consistency. After 10 minutes, perhaps longer, depending on temperatures, there will be a discernible thickening of the soap. If you wish to add no other ingredients, you will have a very serviceable soap.

Keep stirring until the stirring stick will stand alone in the mixture or until the soap is as thick as a good thick gravy. Pour the soap into molds, unless you decide to add scents and other ingredients.

SECONDARY INGREDIENTS IN SOAP

There are several inexpensive and desirable ingredients that can be added to the basic soap recipe. The first of these is a small amount of borax. If you want more suds, add 1/2 cup of borax as soon as you have mixed the lye and lard well. Pour the borax slowly into the soap mixture and stir until it is thoroughly dissolved.

If there are any tiny specks of dark grease and you wish to have these bleached out, you can add a tablespoon of turpentine, although there is likely to be no real reason to add it. Some people like to add 1/2 cup of ammonia, although again there is no real reason to do so unless you want to kill the grease or lard odor, which is not really offensive and which can also be neutralized by the addition of perfumes or scents.

PERFUMING THE SOAP

If you have a favorite perfume or toilet water, you can add 1 teaspoon to the soap, but highly expensive perfume can add significantly to the cost of the soapmaking. A better idea may be to add perfume from a bottle that holds a less desirable scent.

It is not a good idea to add shaving lotions or similar toilet products because the high alcohol content may kill much of the action of the soap.

You can create your own fragrances. It is very simple to do so and in most cases will not cost a cent. If you have a sassafras tree in your area, you can pull off a few leaves or a twig or two—a small portion of a root is much better. Crush these well and immerse in boiling water for several minutes. Keep the sassafras in hot water until you have a dark tea, which can be added to the soap. You can even add the twigs and roots, but you should remove them before you pour the soap into the molds.

Another easy scent to make is a good, rich pine aroma. You can gather several handfuls of pine needles and crush these in your hands by bending, breaking, and rubbing the needles together. Tie them into a small bag or cloth and immerse the cloth and needles in hot water. Allow them to steep until the scent permeates the water well. The water can be added to the soap; if you have a cup of scented water, then add it and cut 1 cup of cold water from the lye mixture.

If you have other favorites, such as flowers, gather several of the flowers, crush them, and steep the flower petals or the entire flow in warm water until the scent is picked up by the water. Magnolia and gardenias have a strong but highly desirable odor.

You can also use cedar, marigolds, wild cherry, and a wide assortment of other natural scents for your soap. You should remember, though, that it takes a good, strong scent to permeate 9 pounds of soap, which is the yield for the recipe given above, so be very generous with your addition of fragrances. Do not expect one rose to scent the entire soap mixture; instead, use a dozen. You can also gather a large bouquet of honeysuckle for one of the finest fragrances on earth.

POURING AND MOLDING THE SOAP

Virtually any sort of shallow cardboard box can be used as a basic mold. First line the box with waxed paper and press the paper down well so that it conforms to the shape of the box or mold. Then pour the soap into the mold. Stationery boxes work well, as do candy boxes and similar containers.

Do not use a deep mold since the soap must set up and cure until it is dry and hard all the way through. A deep mold will not allow proper curing because the center will still be soft while the top will become too hard to cut properly.

For best results, keep the thickness of the mold to no more than 2 inches, and 1 or 1 1/2 inches is more desirable. The width or length of the mold is immaterial as long as the soap will fill the bottom of the container to a proper depth. You would not want soap bars so thin that they would break easily. If you have an assortment of small boxes, such as those that costume jewelry and similar items come in, you may use these for interesting soap bar sizes and shapes.

Let the soap set for at least 24 hours before you cut it. You can watch the setting progression by using the tip of a finger to press the surface of the soap lightly. Within 2 hours the slightest touch will

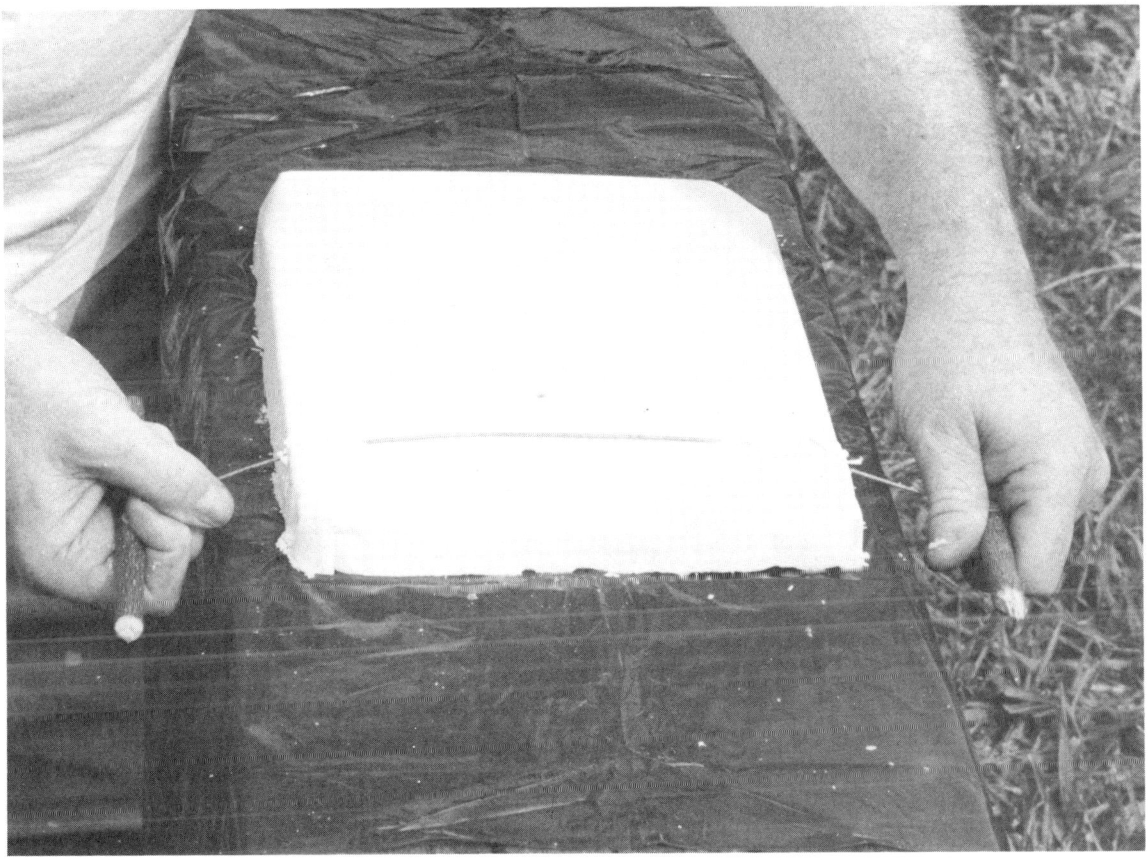

Fig. 19-3. You can use a wire connected to two basic handles to cut your soap. Place the wire on the surface and then pull down steadily with increasing pressure, and the soap will cut easily.

Fig. 19-4. While the soap is still soft, you can mold it or cut it by using cookie cutters or other molds, or you can cut it into geometric shapes. The children will also enjoy carving shapes that have their initials or favorite designs.

leave an indentation; within 6 hours the indentation will not fill and your finger will not have soap clinging to it; within 12 hours there will be a slight hardening; within 24 hours the soap will resist fairly firm pressure. Now is the time to cut the soap into bars or special designs.

CUTTING THE SOAP

If you choose to cut your soap while it is still fairly soft, you can get good results with a thin wire and two 6-inch pieces of small wood. Wrap the ends of the wire firmly around the pieces of wood so that you can hold the two wood handles and let the wire go between your first and middle fingers. Hold the handles so that the wire is very taut and lower it until it touches the soap mold after the soap has been removed from the cardboard boxes. Exert firm, even pressure and the wire will slide through the soap easily and smoothly. Be sure that the soap is on a smooth, flat surface, however, or the mold might break (Fig. 19-3).

If your soap has set up to the point that it is too hard or too brittle for the wire, you can use the point of a pocketknife or kitchen butcher knife to cut the soap into bars. Make the first cut by letting the point extend only a quarter of an inch deep. The next cut in the same groove should be slightly deeper and the next still deeper until the soap divides easily. If you try to cut all the way through the first time you may crack the soap mold or cake.

While the soap is still not set up but has reached a solid consistency you can use cookie cutters or other decorative cutters and cut soap bars or designs in the shape of Christmas trees, stars, bells, geometric shapes, snowmen, apples, hearts, and a multitude of other shapes (Fig. 19-4). You can gather the scraps, grate the larger ones, and use them for washing detergent. Nothing need be wasted.

AGING THE SOAP

After the soap has been cut, allow it to age for as long as is reasonable for your circumstances, but certainly not less than 2 weeks.

The soap will become harder and drier, and the more it dries and hardens, the longer it will last and the more it will lather.

One of our neighbors, who's a veteran soap maker, relates that he had stacked his cut soap on a shelf and somehow one bar fell behind the shelf into some other materials where it lay unnoticed for more than 2 years.

"We used that bar of soap longer than you would believe," the man reported. "It took us months to realize that it was back there, and then we didn't use it for several more months but when we did, it lathered more than any other soap we had ever seen and after weeks and weeks of use the bar didn't seem any smaller than it was when we started using it. I think we may have thrown it away finally; I'm not sure it ever wore out.".

Many soap-makers insist that soap should not be used for at least a year, and some even suggest much longer. You may however use soap that has aged for a month or longer.

To age it, you should spread it or stack it in a reasonably cool and dry room and so that air can circulate freely among the bars.

While your soap is aging, you save your grease and ashes for another run. Of course, you will want to use some of your ashes in the garden, but even if you don't have ashes for the lye water, you can still save your grease and buy a 1-dollar can of commercially prepared lye to make your soap.

20

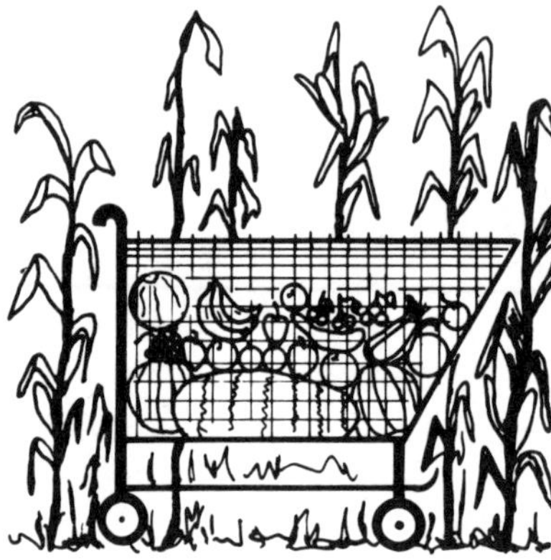

Saving Time, Energy, and Money

One, if not the primary, reason for engaging in home food preservation activities is the possibility of saving time, money, or energy. Admittedly, there are other excellent reasons, including better taste, higher quality, safer methodology, and less salt. Best of all, these measures do not require extensive equipment or complicated techniques.

The dollar has been buying less for the past decade, and most homemakers are increasingly dismayed to learn that one single commodity in the supermarket may have increased in price as much as 50 percent in a few weeks. Most discouraging is the fact that wages may have increased by as little as 5 percent in a given year.

Many consumers have attempted to limit the amount they will pay for a particular product. These consumers must eat, however, and eventually they will be forced to concede to the market demands for more money.

There are many ways that you can save a great deal of your gardening money, and the following suggestions have all been tried repeatedly and found to be not only workable but highly satisfactory.

SAVING MONEY IN THE ORCHARD AND GARDEN

When the prices of foods rose rapidly, so did the prices of seeds, fertilizers, and other gardening necessities. There are many ways to beat the system; however, you can save considerable money simply by shopping for seeds in the grocery store rather than in the garden shops.

Sunflower Seeds

Sunflower seeds, as we have suggested earlier, are delightful in taste, and the flowers themselves attract dozens of beautiful and desirable birds that help in many ways to control garden pests and add beauty. The sunflower seeds can also be used in many ways in your diet, and they provide several desirable vitamins and other nutrients.

Sunflower seeds cost $1.29 for a tiny package, which, means that you are buying 1/3 ounce at $3.87 per ounce or $61.92. per pound. Even though you are buying only a small portion of a pound, the price is still astronomically high. Your package contains; generally 50 seeds; so you are paying almost 3 cents per seed.

Now go to the local grocery store and look at the bird seeds. You can buy a package of sunflower seeds for $.99, and the package contains 2 pounds. If the supermarket package sold at the same rate as the garden store prices, this package would cost more than $120, and no one would think of paying such an exhorbitant price.

While the example is not truly typical in the sense that the typical gardener who plants in volume will buy at a huge discount, and while you will not usually think of planting such an immense amount of sunflowers, the same principle applies to all of your seed purchases. Keep in mind, too, that even if you buy far more seeds than you need, you can plant as many as you like and either feed the excess to the birds, eat them yourself, or save them to feed the birds during the winter.

You can also ask neighbors or fellow gardeners to buy seeds in a cooperative exchange program with you. You can divide the seeds, and no one will need to spend a large amount of money.

Incidentally, the supermarket sunflower seeds germinate in a fine fashion, and the flowers grown are as good as those from seeds purchased from seed stores or catalogs. The same is true of most other purchases from the grocery store, and the savings are always significant.

On the subject of savings, keep in mind that if you order from a seed catalog, you will be expected to add a shipping and handling charge to your initial price, so that you will be paying about 1 dollar extra for each order.

Other Seeds

Check other supermarket prices for your planting needs and you will find that 2 pounds of pinto beans cost about $.39 or little more than $.01 an ounce, while in the garden stores the same beans cost $.29 per ounce. In this instance you may wish to buy a large amount for planting, since a pinto bean takes up considerably more volume and weighs more than a sunflower seed. You will find that 2 pounds, however, will plant several nice rows of beans and leave you with enough beans for a meal or two. Since pinto beans are an excellent source of protein, you can use the beans as meat substitutes, thereby saving a great deal more money.

In the dried bean department of the supermarket you will find lima beans, butterpeas, pole beans, kidney beans, great Northern beans, and many other types that can be planted straight from the package. These beans typically will cost about 10 cents more per 2-pound package than do pinto beans as opposed to much higher prices for seed store beans.

Kidney beans from the seed store cost about $4.60 per pound; pole beans cost about $7.20 per pound, or slightly less, depending on the variety; butterpeas and other peas and beans will typically cost about $.50 per ounce.

Again, buy a package at the grocer's and plant what you need and plan to eat the rest. If you decide to plant seven types of beans or peas, you can save—assuming that you are going to plant one fairly long row of each—almost $20.00 by buying at the supermarket, and you will have seven meals or more from the purchase as a bonus.

While virtually every vegetable in the dried bean section of the supermarket will be successful for planting, you are by no means limited to beans and sunflowers for your purchases. You can also buy a regular bag of potatoes and use some of them for eyes to plant, or you can peel the potatoes so that you have some flesh left on the peelings and plant these for potato hills. Seed potatoes usually cost about $1.00 per pound through seed orders, and sometimes much less from local stores, but a bag of potatoes can be purchased from the supermarket for about $1.50 to $2.00 for a 10-pound bag, depending on the season. In addition to saving $8.00 on the seeds, you also get several meals from the bag.

You can save seeds from a large number of supermarket fruits and vegetables, dry the seeds, and use them in the next planting season. Among the vegetables from which seeds can be saved are tomatoes, okra, mature squash and zucchini, pumpkins, green and red peppers, peanuts, cucumbers, corn, gourds, and popcorn. You can also save seeds from cantaloupes, honeydew melons, watermelons, apples, peaches, apricots, cherries, plums, pears, strawberries, and other seeded fruits.

Keep in mind that many of the items in the grocery store are hybrids and that seeds from these will not reproduce the same kind of crops. Corn, watermelon, and tomatoes, usually are of a hybrid nature, but if you are interested only in good food rather than hybrids, you can still plant the seeds and get a good result.

A hybrid is usually a mixture of two good strains of fruit or vegetable, and while you cannot be certain exactly what you will get, you can be reasonably assured that you will grow a very tasty product that will have many of the desirable qualities of the hybrid.

Sweet Potatoes

When you purchase sweet potatoes for the table, peel thickly and put the peelings into a small container of water. In a few days there will be tiny sprouts visible. When these sprouts grow 6 or 8 inches, remove them from the water and set them out. If you can, cut the peeling so that each sprout will have a section of the parent potato to grow from.

You can put the sweet potato into a box of decayed sawdust and allow it to sprout there. Be sure to keep the potato from freezing, and see that there is enough moisture for the sprouts to maintain their strength. When the weather is warm enough, cut the sprouts from the parent potato, making sure to leave enough potato flesh attached so that the sprout can continue to receive nourishment until it can form its own root system, and then set out the sprouts.

Two sweet potatoes will provide as many sprouts as you can buy for $7.00, typically, in the garden stores.

Rooting Tomato Suckers

You can, of course, start tomatoes from seeds, but the time required is typically fairly long from germination to harvest. Usually from the time you set out the plants, it takes about 70 days for you to harvest tomatoes. The time for ripening then is such that it is very difficult to have more than one crop of tomatoes unless you buy more plants as the season progresses or start seeds at intervals throughout the early spring and summer.

One workable plan is to plant your first seeds early in the year, about February 1 or shortly afterward. You can start the seeds in a jiffy pot or one-step container, but you must keep the pots and the later seedlings indoors until it is warm enough to transplant them. You must have a sunny spot in the house or greenhouse, and you must also be sure to water the plants as they need it.

About 2 weeks after the first planting start a new batch of seedlings. Do the same for three or four plantings, and keep all pots and seedlings indoors until the danger of frost and freeze is past.

Of course, you can simply wait until warm weather and buy your tomato plants. You can buy a dozen or two for less than a dollar if the plants are just root systems and stems wrapped in paper. You will not at this price be likely to find plants with root systems already established in a ready-to-plant pot composed of materials that will blend quickly with the soil. If you buy plants ready to set out, you may expect to pay from 10c to 50c each for the plants, but as the season progresses you will find it increasingly difficult to find plants to purchase.

There is a delightfully quick, highly efficient, and totally free method of having superior tomato plants in a matter of a few days. This method is called sucker rooting.

Once your early tomato plants develop into healthy, strong vines or bushes, you can start to look for suckers. You will find these suckers growing in the middle of a fork in the plant. You will see two strong shoots in the shape of a V and in the center of the V there will be a third shoot that has a nodule or swelling at the base.

You can snap the middle shoot off quickly and easily without damaging the plant. Hold the stem or limb of the plant just below the V with one hand, and with the other hand grasp the middle shoot near the base and bend it to one side until you hear a soft snapping sound. Repeat the process by bending the shoot in the opposite direction. You will hear a second popping sound, almost like the sound made when you break or snap string beans, and the sucker or shoot will be free of the parent plant (Fig. 20-1). If a few plant fibers remain intact, you can clip them with a sharp pocket knife or similar instrument.

To root the shoots or suckers, simply use a small cup or pot with good rich soil—the dark soil that can be found readily in a forest. Put an inch of soil into the bottom of the small pot, stick the end of the sucker into the dirt, and pack dirt firmly around the stem of the sucker all the way to the top of the cup or peat pot.

Fig. 20-1. Find a tomato sucker and snap off the middle shoot. If it is a true sucker it will break off easily as you snap it abruptly in either direction. If for some reason fibers do not break, snip them with a knife.

Place the pots into a shallow pan—an aluminum pie pan will suffice—and water generously. Place it in an area where the suckers will receive plenty of light but no direct sunlight for a day or two. Then move them so that they will receive a good mixture of sunlight and shade for another 2 or 3 days. At this time the suckers are starting to form roots in the soil, and in ten days the plants will be ready to transplant into the garden.

If you don't have peat pots or similar containers handy, you dig a small hole 6 inches in diameter and 6 inches deep. Fill it with good rich soil, preferably peat moss or decayed vegetable matter from a wooded area, and plant the suckers as soon as you break them from the parent plant. Water them thoroughly and if convenient, shade the suckers for a day or so while the first roots are forming.

Be sure to have about 6 inches of the sucker under ground if you plant directly. Do not allow the soil to dry out until the plants have rooted and the new plants are well on their way to a healthy growth.

The beauty of the sucker rooting process is that you are removing unproductive growth from the parent, plant securing new plants at no cost, starting with plants already a foot high thereby shortening the harvest time by several weeks, and making the transplant coincide with your other duties, space, and needs.

You can remove suckers from your tomato plants all season long and start new plants even as late as August and September. You will get the hybrid tomato, too, since you are simply dividing an existing plant instead of starting a new one.

To have late green tomatoes for preserving, start suckers at intervals during the summer so that you will have good green tomatoes to pull just before the first killing frost.

It is unlikely that you will enjoy 100 percent success with your sucker rootings; in fact, a 75 percent rate will be highly satisfactory. You can, however, improve your chances of successful rootings by using the best soil that you can find, by buying a small bag of potting soil or by using one of the special rooting compounds that are available commercially. With the rooting compounds, usually a powder, you simply stick the stem of the sucker into the compound, allow the powder to stick to the stem, and then plant the sucker.

Without anything except patience and care for the plants, you can have a high enough success rate that you can save several dollars on tomato plants and have a very enjoyable time in the process.

Volunteer Plant Gardening

A volunteer plant is any one that you did not plant; yet it somehow appeared in your garden and produced nice vegetables.

How did the plant grow in your garden, uninvited? It came from a seed, that was either dropped accidentally, blown by the wind, passed through the intestinal tract of some bird or animal, or was released from a vegetable that was purposefully or inadvertently left in the garden area.

In one garden plot you may have a row of beans, and during the course of the summer one bean pod fills, dries out, and falls from the plant before you pick it. As the pod decays, the beans are liberated. The following spring as you are tilling the soil you cover the beans with soil, and the beans then sprout. You suddenly have a bean plant or two in the middle of the new cabbage patch, especially if you, like many gardeners, rotate your crops.

You may mulch with table scraps, and in the discarded scraps are seeds from a cantaloupe. You may have missed a small potato when you dug potatoes the previous autumn.

Some birds love tomatoes, and afterward the seeds pass through the birds' intestinal tracts while the birds sit on the fence rail at the edge of the garden. The following spring you will find a bevy of volunteer tomato plants under the fence.

If these plants are a nuisance, it is a simple matter to chop them with a hoe. On the other hand, you may decide that you want to keep the plants but that you need to transplant them.

If you transplant, dig the holes where you want to reset the plants and have everything ready. Then you can use a trowel, a small shovel, a hoe blade, or post hole diggers to remove the plants.

If possible, choose a time shortly after a soaking rainfall to move plants. While the soil is wet it will stick together well, and you can move the root ball and not really disturb the plant.

Insert the point of the trowel into the soil about 4 inches from the base of the plant, if the plant is rather small, and about 6 or 7 inches away if the plant is over a foot high. Run the point into the soil straight down. When it is 5 or 6 inches deep, tilt the handle away from the plant base and pry up. If the soil is damp and soft you will lift the root system gently from the soil, and, holding the root ball together with your free hand, carry the plant to the hole you have already prepared.

Set the root ball gently into the hole and pack soil around it. Pack soil tightly enough so that the root fibers remain in contact with soil but not so tightly that water will have difficulty seeping into the root system. Once the plant is in place and the soil is packed, you can cover the disturbed soil with a thick layer of mulch to hold in moisture.

If the transplant is a tomato plant, you may wish to slide a stick slightly smaller than a pencil against the stem and push the stick into the soil to prevent cutworms from gnawing the tomato plant down.

You can transplant a wide variety of volunteer plants, including tomato, potato, watermelon, corn, pumpkin, cantaloupe, if you move them before the root system has become too well developed. After the plant has been moved, make certain that the soil around the plant stays moist constantly until the plant has adapted to its new environment.

Of course, if the plant is not a nuisance where it came up, you can leave it. There is really nothing wrong with having a huge pumpkin plant growing in your corn field. The pumpkin or similar ground-covering vine will steal very little nourishment from the corn or other plants, and will at the same time prevent grass and weeds from growing, since the broad leaves will cut off direct sunlight.

You would not want the volunteer plant to engulf a lettuce bed, however, and it should be transplanted while the volunteer is young enough to be moved safely. If a stray stalk of corn "volunteers" to grow in the bean row, so much the better. The corn will provide a climbing pole for the beans and the beans will stay off the ground and you will not lose as many to fungus or decay.

Letting Beds Reseed Themselves

While you can save money by keeping volunteer plants, you can also decide what plants will "volunteer" in your garden or other spots on your property. For instance, if you have the space, you can start a mulch pile away from the house and dump all your table and garden refuse into the compost heap. Be sure to dump all types of seeds in the compost heap. With a little practice, you can tell quickly what sorts of seeds have sprouted, and you can have access to many types of plants and even trees from your discarded table scraps. In addition, of course, you get a lot of rich compost for your garden.

There are other free gifts from the garden that require little or no work. Many plants reseed themselves every year, and all you have to do is keep weeds down and occasionally thin the bed and apply a little compost or fertilizer.

If you choose a special portion of your garden, usually an upper or lower strip about 3 feet wide, and plant segments of the strip in lettuce, collards, turnips, mustard, rape, kale, spinach, chard, radishes, cherry tomatoes, and similar plants, you have established a perpetual garden, once you have achieved one full stand or crop. Onions, too, will self-perpetuate themselves readily.

All you need to do, once the peak of the crop has passed, is to allow the plants to "go to seed." In this process the plant, lettuce, for instance, will complete its growth cycle, and after producing abundant greenery for salads and sandwiches the plant will suddenly shoot upward in a remarkable growth spurt. It will become leggy and slender, with leaf clusters protruding awkwardly from the stalk. Shortly afterward seed pods will follow tiny blossoms, and after the seed pods become dry and brittle they will pop open and release hundreds of tiny seeds, which fall onto the soil. Many of these seeds will lie dormant for only a few days before germinating and taking root. A lettuce crop that was planted in the earliest spring weather will complete its cycle and start another in mid-June or earlier. By mid-July or sooner you will have a second crop without any work on your part, and a third growth will develop in August and will continue until frost kills off the parent plants. When winter destroys the chlorophyl growth, the seeds dropped from the last pods will go into a dormant state and will lie just under the soil until the spring warmth renews the cycle.

Lettuce, kale, spinach, chard, and other plants are far hardier than many people realize, and it is not uncommon to have fresh lettuce in very cold weather. If you set a few cement blocks and stretch a sheet of plastic over the patch, you can have winter lettuce until snowfall. In warmer areas is it possible to grow greens all year, even through moderate snows.

Helping Plants To Recycle

There are other plants that, while they will not reseed themselves the way greens will, will continue to reproduce with a little help from you. If you decide to start a vineyard and buy a couple of grape vines, you can turn the pair into a dozen vines in a very short time simply by letting one branch drop from the support wires to the soil, preferably 6 feet or so from the parent vine. As the tips of the vine touch the soil, tendrils will immediately seek a hold. Quickly the plant will send roots into the soil, and a new vine is born.

Before the roots are too deep for you to move the plant, you can clip the vine 3 feet or so above the ground level, dig up the root ball, and move the vine to whatever location you prefer. There it will exist as an independent growth. Keep repeating the process, and you can convert your initial vines into as many as you can use.

Earlier it was suggested that you throw your seeds into one area of the compost pile. You can collect dozens of fruit trees in this fashion and save a great deal of money. When you see seedlings appear, mark them so you will be careful not to dump more compost on them and kill them, and wait until a good time to make a transplant.

To transplant the seedlings, if you can dig up the root ball, carry the entire root ball and seedling to a predug hole. Pack good soil around it and cover the disturbed earth with mulch to hold in moisture.

One good method is to dig a hole, remove the red dirt, and fill the hole halfway with good compost or woods dirt. Then pour in a gallon or so of water and mix until you have a thick soup of earth and water. Set the root ball into the soup, pack dirt around, and mound the area slightly before mulching.

Again, you will not get hybrid fruits, but you will get good fruits. It's a matter of whether you want the hybrids with their obvious attributes or whether you want to save money. A good apple, peach, or pear tree will cost about $15.00, perhaps slightly less, from a nursery. A seedling will cost less than 5 cents, plus your time.

Of course, if you have an apple seedling, you can graft onto the tree whatever type of apple you wish to have, with a few minor exceptions. You can also bud graft peach trees as well as other types of fruits. In certain instances you can even have a single tree with several types of apples on it or a variety of fruits that are compatible with each other on the same tree.

If you don't have any hybrid trees from which to take the bud you can go to a local orchard around pruning time, and the owner will in all probability give you access to all of the clippings you want. If it is not near pruning time you can still find fruit growers who will snip off a limb end for you to use.

When your squash and cucumber plants start to bear, allow one growth to remain on the plant until it is very mature. Then split the fruit open after removing it from the plant, remove the seeds, and allow them to dry in the sun. You can simply leave the fruit under the plant and allow it to decay. As the seeds are exposed to sunlight, they will dry, then germinate, and small plants will appear in a short time. They can be transplanted quickly and easily.

When you are digging potatoes, leave one potato in the hill and cover it as you complete the digging. The potato will in many instances keep well through the winter and will be ready to start new growth the following spring. Keep in mind that in the early spring it is often too wet or too cold to prepare the soil, and if you plant in the fall, you will be ready in the spring.

If you want to set out strawberries, you can buy an everbearing type which is well worth the money, and then you can help the plants to self-perpetuate themselves. You can buy a package of 25 plants for about $9.00. After the plants are set out and are growing well, they will send out runners that will send down roots and form new strawberry crowns. You can leave the new crowns where they root, if you have plenty of room, or you can dig them up and transplant. If you transplant early enough in the fall for the new crowns to establish a good root system, you will have strawberries the following spring.

If you elect to buy strawberry plants, buy the everbearing variety. Typical strawberry plants will bear for 3 weeks to a month and are usually through bearing by the end of May, but the everbearing variety will continue to produce strawberries throughout the entire summer.

To transplant a new crown, simply snap the tendrils that are connected to the parent plant, insert a shovel or trowel under the root system of the newly formed crown, and lift the root ball from its growth area. Place the root ball into a hole, water well, and cover the earth with mulch so that there is a 2-inch layer of mulch completely surrounding the new crowns.

USING NEARLY EVERYTHING FROM THE GARDEN

When you harvest your crops, no matter how large or small, it is inevitable that there will be some waste. Nature, however, did not include any form of waste, realistically speaking, in her plans, and you will find that you can use virtually everything that comes from your garden.

For instance, when your early peas have completed their growth cycle, you can pull up the plants and use them for mulch, or you can plow or till them back into the soil. Peas are one of the best known sources of nitrogen for your soil, and it is truly a waste to discard the old plants by dumping them into a ditch out of sight.

If you want to build up your soil quickly and well, there are few ways better than to plant the entire garden in soybeans. When the soybean plants are fully mature, you can plow the entire crop into the soil and allow the vegetation to decay.

Be sure to return all corn husks, corncobs, pea and bean hulls, peelings, and all other "garbage" from your garden back to the earth when you have finished preparing your food for preservation.

We heartily disapprove of spreading chemically or commercially prepared fertilizers all over your garden. For one thing, the fertilizers are too strong for the soil in many cases, and earthworms and other helpful creatures in the soil are killed almost instantly. For another, commercial fertilizers do not keep the soil rich, or so it seems. On the other hand, a thorough mulching will enrich the soil and keep it fertile for years.

Because commercial fertilizers are very strong, if you are not careful you will burn the roots of plants so severely that the plants will turn yellow quickly and die, or the fruits will grow so rapidly that they are mealy, pulpy, tasteless, or often strong

This is not to say that there is no place for chemical or commercial fertilizers; assuredly there is. Immense farms could not operate without fertilizers, but a backyard garden can be maintained easily and well on mulch alone.

Our greatest argument, perhaps, concerning the commercial fertilizers is that often the fertilizer is loaded with weed seeds. When your row crops come up, they will suddenly be swamped by weeds, usually all of one kind, and you will spend a large part of the season pulling weeds and hoeing. This time could well be used in more profitable activities.

Large-scale farmers, of course, can also use a weed killer as they plant, and they do not experience the weed problem; however we also harbor some concern about the numbers and amounts of poisons that are released into the soil. While we are reassured by various governmental agencies that there is no danger to the environment, we feel that the best course of action is one that has been known to be safe for thousands of years.

Waste As Mulch

Briefly, we mulch and we recommend mulching to all gardeners. What if you don't have a ready supply of mulching or compost materials? First of all, you may have more than you realize. After you have used all of the waste from the garden, try the lawn. After every mowing rake up a huge pile of grass clippings and spread them in your garden. One of the best possible mulching approaches is to plant a row, then spread a thick layer of grass or other mulch materials beside the row so that you have a walkway of sorts. This walk will allow you to pick your crops easily without packing the dirt or miring in soft, moist soil. It will also allow you to garden without worrying much about weeds, because weeds have a difficult time invading the garden if there is a thick layer of mulch to shut out sunlight.

After the garden season is over, plow or till the mulch under and let it decay and form rich soil for next year's garden.

If your grass clippings and garden mulch are not sufficient, use leaves from your trees. Rake leaves and spread them around the garden, and if you still don't have enough, use sawdust, from your own firewood. Try to saw in one spot so that you can collect the sawdust easily and use it for mulch or compost. Wait until it is inert before you use it; green sawdust can have a deleterious effect on some plants if it is placed too close to root systems.

If you live in a city or town, call the city hall to learn about the leaves that are collected each fall. In many localities anyone who wants the leaves and will haul them may have tons of the valuable mulch. If your town has such a policy and if you have a truck, spend a few hours getting the leaves transported to your garden. In many instances the leaves are already in large plastic bags and loading and unloading time is negligible.

If all else fails, you can always use your daily newspapers for weed control. When the plants are up and readily identified, spread newspapers under and around the plants or along rows so that the newspaper sheet is only an inch or so from the base of the plants. You may want to throw some dirt on the newspapers to hold them down. The newspapers will prevent weeds from invading the garden, and the papers themselves will decay and add to your soil. Best of all, they will hold moisture in the soil so that even in dry periods of the summer your plants will have adequate moisture.

We do not recommend the use of newspapers as much as we prefer organic mulch. If you are worried about the ink used in the newspapers, you can rest assured that the tiny amount of ink that was needed to print the papers will do no real damage to plants or soil.

If you have evergreen trees available for your use, rake the pine needles and use these for mulch or compost. Pine needles are superior for use in the garden since they decay quickly and add a great richness to the soil. They are also not broad enough to present difficulties when you are trying to plant or till.

Other Uses

Waste can be used for more than mulch. You can use virtually all of many crops.

If you have a good pumpkin harvest, remember that you can use the pumpkin for pies and other dishes. The pumpkin seeds when boiled and dried are delicious, and you can get a fine supply from one pumpkin. When you scrape the insides of the pumpkin out, separate the seeds from pulp, wash them well, and then cook them in a pot of boiling water for several minutes. You can then either toast them in the oven or spread them in the sun to dry, or you can leave them in the dehydrator for several minutes, until the moisture has evaporated and the seeds have lost their slimy taste. Then crack them and eat the kernels just as you would a sunflower seed or nuts.

With watermelons, eat the traditional part or dehydrate it for later use, and then use the green part of the rind for a fine pickle. Feed the rest of the watermelon to your chickens or use the rinds as compost.

When working with apples, remember that you can dehydrate the apple slices, or you can cut off the meat and cook it for applesauce or apple butter, and then you can use the peelings for making apple jelly. You can also pickle apple slices. You can save the cores or seed sections to dump into your compost heap and use the sprouts that will appear the following year.

SAVING ENERGY IN FOOD PREPARATION

When you are preparing food, there are a few ways in which you can conserve time, energy, and money. If you can prepare several items at the same time or at least use the oven or burner heat, you can save considerably. For instance, if you need to bake biscuits and a cake in the same morning, remember that the cake will bake at about 325 to 350 degrees and the biscuits will bake at 425 to 450 degrees usually. The biscuits, however, will bake for only 10 or 12 minutes while the cake may require 1/2 hour or more.

Turn on the oven and set it for the proper temperature for bread, and while the oven is heating you can mix the dough and roll out the biscuits. It takes about 5 minutes or so to prepare dough and cut the biscuits, and in that time the oven can heat to about the proper temperature.

While the biscuits are baking, mix the batter for the cake. A plain white layer cake can be mixed easily in 10 minutes, so when the biscuits are half baked, reduce the oven heat to the proper setting for a cake. The oven will hold the heat needed to complete baking the biscuits, and when the bread is taken out of the oven, you can put the cake layers in.

If you need to use the oven to bake a casserole or cobbler for lunch, when the cake layers are removed and the oven is still warm, put in the other items that should be warmed. By using basic judgment you can cut your electric or gas bill significantly, since the greatest consumption of energy is needed to thrust the temperature from room heat to over 400 degrees in a few minutes. Less energy is required to maintain the heat once the proper state is attained.

Use the same basic premise with burners on

the range, too. If you need to sterilize jars in boiling water, when the proper sterilization time has elapsed and you are ready to remove the container from the burner, consider whether you need the heat for other purposes.

You could start the boiling water for tea or coffee, or you could start frying meat, with the heat already in the burner. It takes much longer and costs more if you turn off the burner and then 15 minutes later return to cook.

Remember, too, that you can cook several items at the same time, if they are compatible with the same heat level. You can bake potatoes while other dishes are being prepared in the oven, and you can steam certain vegetables while you are heating water for other purposes.

If you plan to fry frozen meat, you can put the meat while it is still frozen solid into a frying pan and cook it. Meat will, however, fry much faster if it is thawed; so you can conserve time and energy by allowing the meat to remain at room temperature for 1/2 hour before you prepare it.

If you have an ice maker, and if you like homemade ice cream, you can put the ice cubes in a larger container and put them in the chest freezer. Let the ice maker continue to make cubes until you have produced enough for the ice cream freezer. Then use the cubes instead of purchasing a bag of crushed ice. The cost of the ice plus the cost of driving your car, which is between 25 and 30 cents a mile for the typical driver, can be offset by freezing the ice cubes and saving them.

You can conserve energy and time, too, by using good judgment concerning your freezer and refrigerator. First of all, if you open the door of the freezer for only a moment, you allow the cold air to escape and warmer air to enter. The result is a minute amount of thawing that results, in turn, in the formation of frost on the walls of the freezer. If you repeat the process very often it will only be a short time before you must defrost the freezer, a task which requires a considerable amount of time and work. Pack your freezer intelligently and orderly so that you can open the door, select the items you need, and close the door quickly.

Plan several days in advance. If you know that you will be preparing a roast on Monday, a blackberry cobbler on Tuesday, and a chicken on Wednesday, then you can take out of the freezer the roast, chicken, and blackberries at one time and avoid opening the freezer door for 3 days. The blackberries and chicken will remain frozen in the refrigerator long enough so that you don't have to worry about spoilage.

If you own a small, portable electric oven, you can use it for broiling steaks, hamburgers, frankfurters, and similar foods. It will also bake biscuits, small casseroles, small pies and in general almost anything your range oven will do but on a smaller scale. Naturally you cannot bake a whole turkey in one, but you can bake chicken that has been cut up and you can bake a turkey breast in some of the larger units. The smaller appliance will use much less energy, and it will not heat the entire house and overwork the air conditioning units.

Remember, too, to allow hot foods to cool completely before you put them into the freezer, since warm foods will cause thawing and frosting.

You can bake many foods in an electric frying pan, too. You can prepare fine biscuits, turnovers, and similar foods quickly and inexpensively in these appliances, and you can even use a covered cast iron frying pan for baking biscuits and other easily baked dishes.

CONSERVING SPACE

We have already discussed how you can conserve space in the food storage area by dehydrating rather than canning and freezing, and one of the greatest attributes of dehydration is that you do not use power to prepare or to preserve the food, except for the small amount of electricity used in a commercial dehydrator. If you use solar power, even this small cost is eliminated. You can then keep your freezer and pantry space open for items that are better frozen or canned than some of the dehydrated items are.

We have mentioned that in the garden you can conserve soil space by planting judiciously. You can let cucumber vines climb garden fences or trellises rather than trail all over the garden and occupy much-needed space. Remember that you can use old automobile tires for your potato patch, and as the potatoes grow, you can gradually fill in around the plants and then, as needed, add another tire, then another, until you have six or seven tires stacked one on top of the other, if your plants continue to grow well. Even if the plants achieve only a moderate growth, you can stack two or three tires and triple your potato production, if you use really good soil, and if you have a good growing season.

We have never had really great success with this method, but we have heard from people who insist that it is the greatest of all methods to grow potatoes. We still feel that a good row of plants, well cared for, is the best way, but if you need the space, it might be worth the time and effort to try the method.

One space saver that truly works beautifully is that of making a tepee of poles. Plant beans all around the base of the tepee, so that the beans can climb as high as the pole tops, and you can pick while standing rather than kneeling or bending. The best part of this approach, though, is that you can grow a large amount of beans in a small circle.

Another excellent space saver is a mini-trellis for beans. You will need a 4-by-4 post or similar sized sapling about 7 feet long, a 1-by-1 or 1-by-2 post 3 feet long, four 1-by-2 posts 3 1/2 feet long, and two 1-by-2 or 1-by-1 posts 3 or 3 1/2 feet long. If you want smaller mini-trellises you can shorten the lengths accordingly.

Lay two 1-by-2 posts parallel and 3 feet apart. At either end nail a 1-by-1 or 1-by-2, then move in a foot or so and nail another at each end. In the exact center nail the shorter 1-by-1 or 1-by-2 flat against the parallel 1-by-2 boards. When you are finished with this step you should have five rails running across the 2 parallel pieces, and the center rail should be situated so that it can be nailed easily to the top of the post or sapling.

Now nail the top piece in position so that the rails are flat and parallel to the ground. Then dig a hole 1 1/2 feet deep, put the post in it, and pack dirt around the post until it is securely positioned.

Tie sturdy strings to the rails—as many as you like but certainly 1 at each end of each rail—and then plant beans at the point where the string touches the ground. You can use a small stick to anchor the string in the soft dirt.

Now the beans can grow straight up the trellis, and you will have 10 more bean plants producing in the space that under usual conditions would have been occupied by one or two plants (Fig. 20-2). Since the beans grow upward rather than spreading out, you will save space and find it very easy to pick the beans.

If you don't wish to go to the trouble of building the bean trellises, you can plant your beans in the traditional rows. At the end of each row, and at 6 foot intervals along the row, put up 4 foot long stakes. They can be cut from old tree limbs, scrap wood, or whatever is handy. Now simply run sturdy cord from stake to stake, and make three strands of cord 1/3 the way up the stakes, 2/3 the way up, and at the top. If cord is scarce, use old tree limbs again and tie the limbs or wood strips in position.

PROTECTING YOUR GARDEN FROM INSECTS

Home food preservation and energy conservation may not seem to be related to the problem of protecting your garden from the various insects that are attracted to a vegetable patch, but if you have ever been disappointed and frustrated at seeing a garden riddled by pests plus needing to replant the garden, you will see the relationship.

There are many types of garden pests, far too many to discuss in anything but generalized comments here. Among the worst offenders are Japanese beetles, bean beetles, tomato hornworms, squash bugs, and potato beetles.

The Japanese beetles are such a problem that methods of dealing with them will be discussed later in this section, but the others are less destructive or at least less universal in their tastes.

You can, for instance, handle the problems of potato beetles fairly easily just by walking through your potato patch and using your fingers to flip the beetles into a jar. Dump the beetles into a container of water, where they will drown, and you have no further problems with them until the next cycle hatches. Every 3 or 4 days you can stroll through

Fig. 20-2. A mini-trellis for bean plants will permit numerous plants to grow in a very small space, and the beans will grow high enough that they can be picked with ease.

the patch and clean the potato plants. The entire operation, unless you have an enormous patch, will take only minutes.

Birds eat insects and will be poisoned if they eat bugs that have devoured poison. If you refrain from spraying, unless it is imperative to do so, you will have more songbirds, and the birds will reduce the insect population to a very great extent.

You can plant marigolds along your bean rows, and the scent of the marigolds will drive away many types of bugs. You can plant a marigold every 3 or 4 feet and it will grow alongside your bean plants.

For tomato hornworms, your best bet is to hand-pick the pests. These worms are 3 to 4 inches long and over an inch in thickness at times, and they are almost impossible to spot since they are precisely the same color as the tomato plants. If you watch your plants regularly, however, you will notice pellets of manure from the worms, or you will see that tips of limbs have been stripped entirely of their leaves. You may also notice that some immature tomatoes have been nearly devoured, usually on the bottom.

When you see these signs, you can start a close visual examination of the plant, you will see the creature. You may wish to use a pair of sticks to pull the worm free of the plant. Once you have him, smash him so that there is no doubt that he is dead. One such worm will virtually defoliate a huge tomato plant in a few days.

Squash bugs and other similar pests can usually be controlled by hand-picking, especially if you can stop the problem before it becomes serious. Examine the leaves of your squash plants for tiny clusters of eggs. When you find the eggs, which look like tiny brown beads compressed into a 1-inch patch, you can smash the eggs with your fingers and thereby prevent the hatching of dozens of the bugs.

Japanese Beetles

Japanese beetles, unlike other pests, are so voracious and so varied in their diets that they will attack nearly everything in your garden. The only commendable qualities that these insects have are the facts that they destroy morning glory vines in your garden, briars, poison oak, ivy, and sumac, and unwanted plants.

The beetles, however, also eat apples and apple leaves, plum foliage, and the leaves of nearly every fruit tree in the orchard. They eat blackberries and other fruits, and they attack flowers, shrubs, nut trees, and virtually every vegetable.

Sometimes you must use a spray to destroy the beetles, since they are so incredibly numerous. You can shake a grape vine and literally hundreds of the beetles will fall to the ground or fly away. You can destroy many of them as they lie on the ground, but you may be fighting a losing battle.

Another reason for spraying, if all else fails, is that the beetles are so high in the apple trees that you cannot reach them to hand-pick them. Since you cannot permit your crop to be destroyed, some drastic steps must sometimes be taken. One such step is to apply milky spore spray to your orchard and lawn. This spray will kill the larvae of the beetles before they emerge to fly away and devour your garden.

Unfortunately, if you apply the milky spore treatment and your neighbors don't, then their beetles will hatch and come to your garden and orchard. If your orchard and yard are very large, the cost of the spore application will be perhaps too great to be practical.

Making Beetle Traps

One possible and inexpensive way to avoid the cost of the spore application and the damage done by the beetles is to make a homemade Japanese beetle trap. It can be constructed from an old plastic milk jug, an empty salad dressing jar, and a foot or so of twine or wire.

To make the traps, remove the lid from the salad dressing jar and use a sturdy pocketknife or similar instrument to cut a 2-inch X in the lid (Fig. 20-3). When you have done so, bend the 4 flaps downward and fold them flat against the underside of the lid (Fig. 20-4). Next, use the point of your knife and cut the inner portion of the milk jug lid out so that you will have an 0.

At this point you need to cut away part of the milk jug. Starting at a point 1/3 of the way from

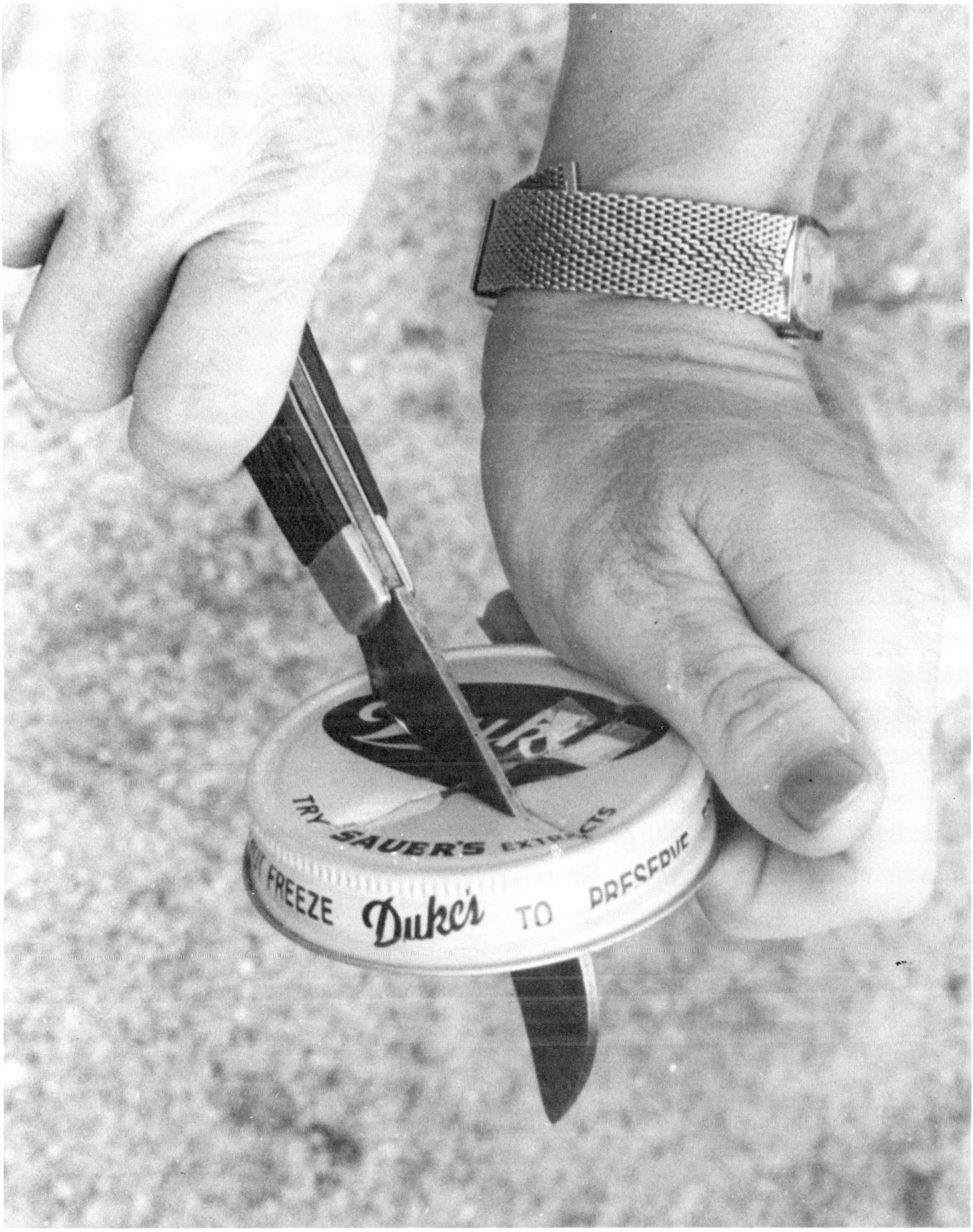

Fig. 20-3. Cut a X in the lid of a jar, and make sure that the cut is wide enough so that you can fold back the flaps later.

Fig. 20-4. Fold the flaps back to the inside of the lid. Keep the opening large enough that the neck of a plastic milk jug can be placed through it.

Fig. 20-5. Cut the inside out of the top of a milk jug so that only a circle remains. After you insert the neck of the jug through the hole in the lid, replace the plastic lid on the milk jug so that the milk jug will not fall. Then you can cut out the sides of the milk jug and screw the metal lid back into the jar you took it from.

the bottom and 1/3 of the way from the handle, cut 2/3 of the way around the jug (Fig. 20-5). Then cut upward on both sides and across the middle and remove all of the section. The top of the jug, one third of the sides including the handle, and the bottom will be left.

Now put the top of the milk jug through the hole in the lid of the salad dressing jar, starting through from the top of the lid. When the neck protrudes, screw the milk jug lid back on. It should now be impossible for the milk jug to be pulled back through the lid.

Then screw the salad dressing jar into the lid. At this point the two containers are attached firmly.

Make two holes about 4 or 5 inches apart in the bottom of the milk jug. Into these holes you can insert twine or wire for hanging the trap in a tree (Fig. 20-6).

When hanging, the milk jug will be upside down with the handle area serving as the rear wall, and the salad dressing jar will be right-side up.

In the space created where the handle of the milk jug fuses with the rest of the container there will be a shallow pocket where you can put bait to attract the beetles. This bait can be a commercial preparation or you can use pulp of fruit. Apples and peaches have a strong odor to attract the beetles, which will fly into the container, smash into the back wall, slide down through the hole, and land in the salad dressing jar. The beetles are then unable to escape and will die of starvation or by drowning, if you fill the jar partially with water.

In a few hours you can catch hundreds of the insects in these homemade traps.

Fig. 20-6. Attach a cord or wire through holes in the bottom of the milk jug, and then hang the beetle trap in a tree. Beetles fall into the trap and cannot escape. No poisons of any sort are used and yet the beetles are controlled well.

SOME FINAL HINTS

While you are trying to encourage your plants to grow, don't overlook some commonplace ideas that really work. For instance, if you cut your own hair, dump the hair around the plants. Since human hair is largely protein, the protein is quickly absorbed by the soil and will be used by the plants.

Another suggestion, researched and applied first by Dr. Douglas Boyce of Gastonia, North Carolina, is that of feeding your plants vitamins. When Dr. Boyce was involved in research during World War II he began to wonder if vitamins would not be as good for plants as they are for human beings. The results of his experiments were rewarding. Dr. Boyce found out that one tablet of

Vitamin A or Vitamin B dissolved in a gallon of water would produce an astonishing growth in plants. His suggestion was that each plant should have 2 ounces of the mixture every week or 10 days. He reports that, in repeated experiments, his vitamin-fed plants invariably outgrew and outproduced other plants in the same soil and in the same environment.

A third suggestion is that, if your garden demands spray, do not spray in the middle of the day. Wait until dusk so that the bees, which are indispensable for a garden's cross-pollination, have finished their work. The poison will have lost enough of its strength by the following morning, if you mixed it according to directions, that the bees will not be killed by it and the nectar used in the making of honey will not be laden with poison.

Please remember that there are instances in which people have been seriously poisoned by honey made from lethal nectar, some of which is natural, as in the case of the rhododendron and laurel, and some of which is the result of sprays that were carried into the hive.

If you have strawberry plants or cherry trees, you can build or create an effective scarecrow by simply laying a length of old hose along the ground or along a tree limb. The birds that ravage your crops will be frightened into assuming that the hose is a snake waiting for its meal and will, unless they are terribly desirous of fresh fruit, leave your crops alone.

You can make other types of bird scarers by simply driving a stake into the soft soil in a row and then attaching an aluminum pie pan to the stake by means of a short piece of twine. When the wind bangs the pans against the stake, birds and rabbits tend to abandon the garden.

If you want to get plants that are frost-resistant started rapidly outdoors, plant the cabbage or other plants or set them out so you can put a cement block over them. Let the plant stick up through the holes in the block. During the day the sun will warm the block and the block will hold the warmth long after sunset, giving the new plant the benefit of extra hours of heat. The block will keep hungry rabbits from eating the plants.

21

Fireplace and Woodstove Cooking

When oil and gas prices rose so sharply in the early seventies, thousands of Americans started looking anxiously for alternate ways of heating. Those who had bought kerosene or fuel oil for less than 20 cents a gallon were horrified to realize that they would now be forced to pay well over a dollar a gallon.

There was something that could be done concerning home heating, however, and that was to return to the wood-burning stove.

Many people who had earlier bricked up fireplaces knocked the bricks out, cleaned out chimneys, and prepared to supplement the furnace's work with the fireplace.

The woodstove and fireplace left a great deal to be desired, however, at least in the mind of many people. Even though the stove and fireplace helped cut heat bills, it was still necessary to use gas or electricity to cook meals. Regrettably, many people still think that the fireplace and woodstove are good only for heating.

The good news is that many foods can be prepared on a woodstove or in or in front of the fireplace; in fact, entire meals can be prepared independently of the kitchen range. Fireplace or woodstove cooking will not require any special equipment other than a swinging metal arm, and you can get by without even this arm if you insist.

First of all, you can cook steaks, hamburgers, spaghetti, soups, cakes, biscuits, baked potatoes, fish, pancakes, vegetables of all sorts—in short, virtually anything, except huge dishes like roasted turkey and similarly complex preparations—in the fireplace.

What special types of firewood are needed? Any of the hardwoods work beautifully, and oak, hickory, and poplar are especially good to use. Poplar, however, has a tendency to crack and pop, and at times huge embers may explode into the middle of the room unless you use a fire screen. Oak and hickory are, by all estimates, excellent woods to use. Pine, unfortunately, blazes too high, and the smell mixes with the food and produces an undesirable aftertaste.

Remember that the charcoal grill, which is

used by millions of Americans, is actually cooking over wood, except that the wood has been commercially prepared—and is therefore very expensive, comparatively speaking—and has been ignited with petroleum products or other chemically suspect preparations.

The charcoal itself is not apparently a problem; however, the danger from the lighter and the subsequent fumes must be viewed with considerable question. There are medical studies that link the inhaling of gas fumes to cancer, and since charcoal lighter is composed of petroleum and naphtha, the inhalation of the fumes is inevitable. The same fumes are passed into the cooking food, unless you wait until the lighter is totally burned away. Even after prolonged burning, though, the taste of the lighter can be detected in the cooked food. You can certainly use charcoal in your fireplace, however, and you can, if you wish to experiment, make your own charcoal so that you will know exactly what is in it.

MAKING YOUR OWN CHARCOAL

Charcoal, in its simplest sense, is a combustible material that has been created in a partial vacuum. The wood has been ignited, and then the air has been cut off so that the wood slowly is extinguish. A dark, carbonish residue is left covering the wood and, in most cases, permeating it.

Commercial charcoal is made, in some instances, at least, in special ovens where complete control over air and temperature can be exercised. If you want to make your own charcoal, however, you can do so without any special equipment.

If you have ever reused charcoal, you have seen one of the controlling principles at work. New charcoal, after it has been used for cooking a batch of hamburgers, for instance, can be dumped into a metal container that has a tight fitting lid. The charcoal uses up the available oxygen, and then the smouldering fire goes out. The next time you use the charcoal it lights much easier and burns to a whitish consistency much faster than it did the first time. The carbonization process is illustrated adequately here; the same carbon that can interfere with your car's ignition system also interferes with the ability of the wood to burn freely and well.

To make your own charcoal, you need a type of wood that you would like to use in cooking. The next time you are cutting firewood you will notice that the oak or hickory tree will have numerous small limbs that cannot be used for firewood. These limbs may have a diameter of 2 or 3 inches at most and are apparently worthless to you. Instead of leaving the limbs in the woods to decay and impede the growth of seedlings, however, cut several of them and take them home. Then, when you have leisure time, use your chainsaw or bowsaw to cut off slices 2 inches thick. Spread these slices in a hot dry place so that they can cure or dry out well.

After the slices are dried thoroughly—it will not take more than a couple of weeks for the summer sun to dry out thin slices of wood—you can store the slices in a dry place until it can continue the process. You can start the charcoal in an outside fire or wait until you have a fire in the fireplace. Drop the slices into hot coals and wait until the slices are burning brightly (Fig. 21-1). Use a pair of tongs to retrieve the slices from the fire and drop the burning sections into a heavy metal container that can be closed tightly (Fig. 21-3).

When the slices are in the container, carry it outside and store it where there will be no danger of a fire and allow the slices to burn themselves out. It works better if the bucket or other container is nearly full so that there will not be enough oxygen for the slices to burn very long. When the charcoal is totally extinguished and has cooled, it is ready to use.

If you choose to do the initial burning outside, one good way is to dig a hole 2 feet square and build a fire in it. You can use pine for the first stage of the fire, and when the fire is advanced to the stage that there is a good bed of coals, spread the coals evenly and drop in the oak or hickory slices, let them begin to burn, and then shovel dirt over the fire. Be sure to shovel several inches of dirt over the coals so that the oxygen supply will be cut off completely. Several hours later or the next day you can shovel out the dirt and retrieve your charcoal, which would then be ready to use (Fig. 21-2).

Fig. 21-1. You can make a good supply of charcoal on the smaller grills by building a fire and waiting until you have a good bed of coals. Stack the cut wood sections over the coals and wait until the wood has been charred thoroughly.

Fig. 21-2. After you have charred the wood and buried it until the fire was extinguished, dig up the cooled nuggets and place them in a large metal container. An old garbage can works well.

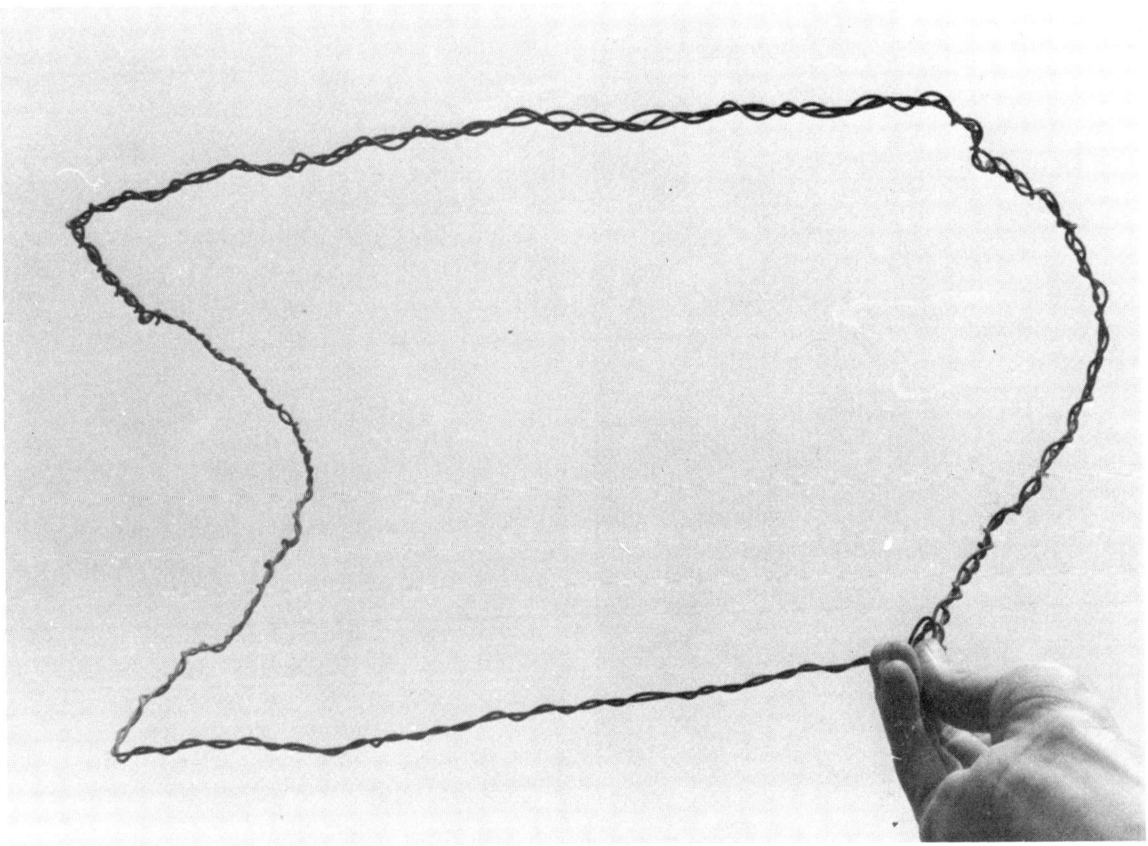

Fig. 21-3. Use wire to form the frame for the reflector oven. You can shape the wire easily by hand so that it resembles an open mouth.

BAKING BREAD IN THE FIREPLACE

When you decide to try your hand at fireplace cooking, a good place to start is with bread, which can be baked easily and well in a makeshift reflector oven or under the coals.

Reflector Oven

If you want to try the reflector oven, you will need several feet of heavy wire—coat hangers will work perfectly—and some wide aluminum foil. If you have a long length of wire you will have an easier time of the frame construction, but if you only have short lengths, bend the ends together and twist them so that you have an uninterrupted length.

Start with one end in your left hand. With your right hand holding the center portion of the wire, make a bend about 15 inches from the end. You will need an acute angle so that the wire in your right hand will bend back over the left hand and about 10 to 12 inches above the first section.

Now bend the top section at right angles just above the end of the bottom section. This new bend should give you a straight section 15 inches long.

Bend the wire again so that you create another acute angle exactly like the first one. At this point you should have two angles, two sides of a triangle, connected with the 15-inch section across the top. Now bend the wire one final time so that it runs from the bottom of the second angle to the bottom of the first one, and then wrap the wire around the end that was initially held in your left hand (Fig. 21-3).

The entire process takes less than a minute,

and when you are through you can cover the framework with aluminum foil, but you must leave the front open. Cover the bottom and the two ends, and your oven is complete and ready to use (Fig. 21-4).

When you are ready to bake bread, prepare your favorite biscuit recipe and spread the biscuits on a cookie sheet or put them in an aluminum pie pan. Your fire should be hot and should have a deep bed of coals. In nearly all cases it is the hot coals that do the cooking, not the flames, so wait until the coals are radiating intense heat. Place the oven about 2 feet from the coals—the exact distance depends on many variables but primarily on the heat intensity of the coals. Put the pan of biscuits inside.

Baking time again depends on the heat from the coals and the distance of the oven from the coals. You will need to make several checks as the biscuits start to bake, but generally fireplace biscuits take less time to bake than do oven breads. Check after 5 minutes, and if the biscuits are browning too fast, move the oven back. If they are not rising at all, move the oven closer.

Assuming that you have a fire in the fireplace for heat or enjoyment, you will in all probability keep the fire going for several hours. If you wish, you can bake several pans of biscuits and freeze the extra ones.

Baking Under the Coals

To bake biscuits in the coals, you will need a large cast iron pan with a lid. The lid need not necessarily fit well as long as it will cover the pan to keep ashes

Fig. 21-4. Cover the wire form with aluminum foil so that you have a shape enclosed on all but one side. Leave the front open so that pans of bread can be inserted easily. If aluminum foil is scarce, you can leave the bottom open as well.

Fig. 21-5. If you prefer, you can make a box-like structure for the reflector oven so that it will hold cakes and other larger items for baking.

and coals out. You can get by with an aluminum pie pan and a sheet of aluminum covering it, but if the coals are really hot the bread will burn unless you are very careful.

When you are ready to bake, arrange the biscuits in the cast iron pan and then put the lid or cover in position. Use a poker or fireplace shovel to rake coals back so that you can slide the pan directly under the grate. When you have the pan in place, rake coals up against the pan on the sides and in front. In about 7 minutes, depending on the heat of the coals, you will have biscuits ready to eat (Fig. 21-5).

Biscuits baked in the fireplace are unquestionably delicious. They brown beautifully and have a wonderful texture. Best of all, there is an extremely faint hint of smokiness, but there is not enough wood odor to hurt the taste. On the contrary, the taste is enhanced greatly. Bread baked in the coals has a slightly greater smoke hint than does bread baked in the makeshift reflector oven.

Another Method

A third way to bake bread in the fireplace is simply to put the biscuits in a pan, cover them with a sheet of foil, and place the pan perhaps a foot from the coals. If you wish you may prop the pan on a piece of firewood or a brick in order to determine more uniform distribution of heat.

A little experimentation will quickly help you to decide what the best method for baking bread may be for your particular needs, tastes, and time requirements. If you don't use the fireplace often but rely instead on a wood-burning stove, you can also bake excellent bread on the stove top.

BAKING STOVE-TOP BREAD

The typical woodstove that is used for heating rather than cooking purposes may be used to bake bread in two distinct ways. One way is to bake the bread in a makeshift oven on top of the stove. The second way is to add an oven to the stove pipe.

A Stovepipe Oven

If you have space and don't mind the somewhat archaic look of one of the stovepipe ovens, you can look around in older stores or catalog supply houses and find one of the round ovens that fit onto your stove pipe. These ovens have a pipe fitting on the top and bottom, and in the center there is a round space divided by a bread rack. All around the baking area there is a thin heat convection area designed that the hot air from the stove completely surrounds the baking area and gives fantastic heat distribution.

Such an oven can be bought for probably around $25 and can be installed in minutes. It stays in place all the time as opposed to the portable ovens that you remove from the heat area when the oven is not in use. Some of these ovens that fit into stove pipes have a thermometer built in so that you can see at a glance how hot the oven is; others have no dials and you must guess at heat levels. The ovens can be used for all types of baking. The only disadvantages are that you have limited control over the heat and that the heat does not stay at a specific temperature. It may fluctuate to a great extent, depending on the type of wood being burned and how the fire box is constructed, lined, and loaded.

A Makeshift Oven

The simplest type of temporary baking device, one which costs absolutely nothing in money or time, is the use of inverted pans. After you have mixed the biscuits and cut them, place them in a small pan, such as a pie pan. Now use a metal or foil pie pan and turn it upside down on top of the stove. Set the pan of biscuits on the inverted pan, and then cover the two pans with a large pot or pan, one that does not have wood or plastic handles, since the handles will be in direct contact with the stove top. The inverted pan keeps the biscuits from coming in contact with the direct heat, and the huge pan or pot used to cover biscuits and inverted pan keeps the heat from escaping (Fig. 21-6).

Biscuits baked in such a fashion are very tasty and compare favorably with those baked in a traditional oven, and there is no cost, since you would be operating the stove for heat anyway.

If you don't want to use an aluminum pan for the biscuits, you can use a fairly large sheet of foil and crinkle it closed so that the hot air cannot escape readily, but be sure to leave enough room so that the biscuits can rise.

PREPARING THE COMPLETE MEAL IN A POT

If you don't have a swinging arm for your fireplace, you can install one or have one installed for a very reasonable fee. From this arm you can suspend a pot filled with all the ingredients necessary for a nearly complete meal, or you can cook one course of a dinner in the pot, which must rank among the finest ways ever to prepare certain foods.

We recommend a heavy cast-iron pot with a heavy lid. When the pot is filled 3/4 of the way to the top with dried beans and water, for instance, the pot can be swung over the fire to one of several positions, depending upon whether you want the beans cooked slowly or rapidly. For best results cook over a low fire that has a a thick, hot bed of coals. The coals produce such intense heat that in a matter of a few minutes the water in the pot will be boiling and steam will be forming with sufficient pressure to push up the heavy lid of the pot so that the steam can escape.

The pot lid is so heavy, if you are using a cast-iron lid, that considerable pressure is built up before the steam will lift the lid. In brief, what you have at this point is a rudimentary pressure cooker. The beans, even pinto beans, will be cooked to a very tender consistency in a remarkably short time, and the soup will be rich and dark.

Another delightful meal in a pot is beef stew.

Fig. 21-6. You can cook an entire meal on the top of the wood stove. Bread and cake layers can be baked in the makeshift oven formed by inverting a metal pan and covering it with a large pot.

Chop beef into 1-inch chunks, then pour a quart of water over a pound of meat, and then chop carrots, celery, potatoes, onions, and any other vegetables you prefer and swing the pot over the fire. You may wish to add string beans, lima beans, tomatoes, rice, and other ingredients as the stew simmers. Check the water level regularly to be sure that there is plenty of liquid in the pot (Fig. 21-7).

The pressure-cooking effect is again in progress, and the meat will be cooked to a wonderful tenderness in a short time. As a bonus, when the lid is pushed up so that steam can escape, tiny amounts of smoke will be admitted into the pot so that the entire stew has the smoky taste associated with barbecue and similar foods.

You can prepare spaghetti sauces, soups of all types, and steamed or boiled potatoes and other vegetables on the open fire. Anything that can be prepared by hanging the pot over the fire can be prepared by setting the pot on the top of the woodstove.

You can prepare fish stews, oyster stews, gumbos, hot cereals for breakfast, hot chocolate, hot teas, herb teas, coffee, and a wide assortment of such foods by using the fireplace pot method.

If you make coffee, you may want to measure out the coffee in a small cloth. Tie a knot in the cloth or fasten it before you start the coffee. In this way you can keep the coffee grounds from mixing in with the brew.

OTHER FOODS TO PREPARE

You can bake potatoes by wrapping them in foil and dropping them into the ashes under the grate. Rake back the coals to make a hole large enough to accommodate the potatoes. When the potatoes are

Fig. 21-7. Stews and roasts cooked in a pot swung over the fireplace are easily prepared and provide fine eating.

in position you can push the coals and ashes back up against them. Potatoes prepared in this fashion are delicious and tender, but they cook much faster than they do in an oven. You will need to check on them after about 15 to 20 minutes. Use a fireplace tool or a wooden spoon to press against the foil. If it gives readily, the potato is ready to eat.

French fried potatoes are easily made by putting the oil into a pot and swinging the pot over the coals. When the oil is hot enough, cook the fries as you would on the kitchen range. You can also make potato chips in the same fashion except that you slice the potatoes very thin before frying them in oil until they are crisp.

You can attach a grill to many of the arms in fireplace assemblies. You can load the grill, swing it over the coals, and cook steaks and other foods just as you would on the backyard grill. Use oak or hickory, not softwoods, and cook over coals, not flames. The result will be meat that is much better in flavor than that which is cooked over charcoal.

While you are grilling steaks, you can wrap ears of corn in foil and cook these on the grill or in the ashes, and you can do the same with fish, pork chops, or chicken. You will probably want to grill the chops and chicken over the open coals, however, in order to enjoy the superb smoky taste.

If your fireplace assembly does not have a swinging arm but you want to grill foods, you can purchase 10 or 12 fire bricks. Either mortar these into permanent position on either side of the grate or simply stack the bricks in position when you want to grill. When the bricks are in place, you can rest the grill securely on the two stacks of bricks.

Another treat that can be prepared in the fireplace, either on the suspended grill or in the ashes, is a hamburger-onion-tomato combination. Shape lean ground beef into 4-inch wide hamburgers that are 3/4 of an inch thick. Lay a slice of onion on top of the hamburger pattie. Excellent onions are available for such cooking, among them the famed Vidalia onions that are sweet, flavorful, and highly digestible, and the onions are large enough that one slice will completely cover the hamburger pattie. Add a huge slice of tomato atop the onion. Now wrap the assembly in foil and put it on the grill with the hamburger pattie on the bottom. If the meat is not extra lean you may want to perforate the foil at the bottom so that grease may drip freely.

You can also put the package into the ashes and allow the 3 foods to cook together, but if you do so, do not perforate the bottom foil since ashes may get into the food.

If you like snacks, such as roasted peanuts, you can wrap peanuts in foil and lay the package in front of the hot coal bed, or you can put the package into the ashes if the fire is rather low.

While you do not have quite as much flexibility on the top of the wood-burning stove as you have in the fireplace cooking area, you can still prepare most of the foods listed above on the stove-top oven described earlier. You can also use a frying pan easily on the stove top.

If you want to fry eggs over the fireplace, you can set the frying pan just over the coals—use a pan with a metal handle—and leave it until it is hot. Remove the pan from the heat, drop in the eggs, and the heat held by the pan will be sufficient to fry several eggs adequately. On the stove top you can fry the eggs as you would in the conventional fashion.

Popcorn is easily handled in the fireplace or on the stove top. Simply prepare the oil and popcorn kernels as you would normally and, using a metal-handled pot, place the corn and pot near the hot coals. As soon as the first kernels pop, pull the pot back and shake it vigorously, then wait for more kernels to pop. Continue this process until all kernels are popped.

Don't forget a dish that has become an integral part of American culture: the hot dog. You can cut sticks and have a hot dog roast any night or day that you have a fire in the fireplace. While the franks are cooking, you can warm the rolls in an aluminum pan in front of the coals.

Part of the fun of fireplace cooking is seeing how many kinds of food you can prepare and enjoy, but certainly a vital part of the enjoyment is the awareness that the heat isn't costing you a cent, unless you bought the wood or unless you count your labor in cutting it as part of the cost.

You probably had the fire built for recreational or aesthetic purpose or to cut the heat bill, but the food preparation is a bonus, an added bit of pleasure or convenience.

Don't overlook the by-products of the fireplace, either. You can shovel out the ashes and spread them on your garden. You improve the soil greatly and thereby improve your garden for the coming year.

FINAL CONSIDERATIONS

While the cooking is easy and the variations on preparation of recipes are broad enough to allow a great deal of latitude, there is not a great deal of room for error in the actual cooking activities, namely cutting the wood, storing it, curing it, and building and maintaining a good fire.

Building a Fire

To build a fire that is heat-efficient and aesthetically rewarding, you need good kindling, good hardwood of medium size, fire-holding hardwood, and a good bed of coals.

If you have a grate, use thin strips of pine, cured and dry, or poplar, also cured and dry, and lay the basis for your fire by making an X with the first two strips of kindling. As you progress with the kindling stack, keep crossing the other pieces at the center so that you have a sort of flattened wigwam. The strips of kindling should be such that plenty of air can pass throughout the stack. The air passages are crucial in that fires depend on drafts for their supply of necessary oxygen.

Don't make a huge stack. Generally 8 or 10 pieces of thin kindling will suffice. If you have a large stack, once the fire catches well you will have tongues of flame leaping up the chimney flue, and if there is creosote left from previous fires you are inviting a chimney fire.

If your chimney should catch fire, fire-fighting authorities say that your best step, after first calling the local fire department, is to determine that your roof remains soaking wet. So while you wait for the fire department, hose down the roof and keep hosing until the firemen arrive.

When you have stacked your kindling properly, wad 1/2 sheet of newspaper and shove it under the grate. Many people like to stack kindling on top of the wadded newspaper, but you are building a paper foundation when you do so, while if you put the paper under the grate your fire will have a wood basis, and you won't have the trouble you have when the paper, which of course burns faster than wood, is consumed and the stack of kindling collapses and scatters your fire.

Once the kindling is burning steadily and brightly, you can start to add small pieces of hardwood, which should be no more than an inch in diameter. If the wood is too large, it will not ignite well. After you have added three or four pieces of small hardwood, wait until the new wood is burning well and then add 2 or 3 pieces of hardwood that is 2 to 4 inches in diameter.

Your object should not be to have a fire with high flames, for the heat comes from the coals primarily, not from the flames. While pine and other soft woods burn with a high flame, they produce very low heat, while the oaks and other hardwoods that produce the great amount of heat burn with very low flames.

Once you become adept at stacking a fire, you will be able to create a fire that has an amazing variety of patterns and colors, and one that is beautiful but not dangerous.

Some Precautions

If you plan to cook over your fireplace fire, add your larger wood at least 1/2 hour before you put on the steaks, since you will want the flames to die largely before you put the meat on. You don't want the flames to touch the meat, since the fire is carbonizing and the meat will turn blackish and have a burned taste. On the other hand, the meat cooked over the coals will have a pure, naturally roasted or barbecued taste that is free of the carbon smell and taste.

When you go to bed, be sure that the fire is properly banked or smothered, and be certain that the fire screen is in place. If you don't have a fire screen, buy one. It is extremely hazardous to go to sleep with hot coals in the fireplace, since there is always the danger that a coal will explode and send sparks and embers into the room. It is not at all uncommon for some kinds of woods to explode embers that will travel 8 to 10 feet.

If you have several large pieces of wood still burning well when you are ready to retire for the day, you can use fire tongs and lift the pieces from the grate and stack them against the back wall of the chimney and then cover them with shovelfuls of ashes. The ashes will keep sparks from escaping and will also hold the heat in and keep the air out so that the next morning you will have a small bed of coals with which to start the day's fire.

When you use a wood-burning stove, of course, you can pack the stove well at night, and you will have a nice fire ready the following morning. You should turn the damper controls down to an almost closed position when you retire so that the fire will not have enough oxygen to burn well, or your fuel will be consumed before morning. A good policy is to turn the damper control knobs all the way closed, then open them 1/4 turn each.

Getting Firewood

If you plan to cut your own firewood, you should cut it at least a year before you plan to use it, if at all possible. Hardwoods, like pines and other softer woods, contain some creosote, which is the culprit in chimney fires, and you can virtually eliminate the creosote problem by burning wood that is dried out and which has been cured for a year. By "dried out" we mean wood with the sap dehydrated, not necessarily wood that has been protected from the rain or snow, although it is certainly a good idea to keep your wood protected from the elements.

If you buy your wood, buy it far enough in advance so that you can allow it to cure. Most woodyards cut and sell wood on a seasonal basis, and it is likely that the wood you buy was cut only a few days before your purchase.

Be duly warned, too, that when you call about a load of wood and are told that you can buy a cord of firewood for a certain amount of money, if the wood arrives in a pickup truck you have not purchased a full cord, unless the truck has sideboards. It is virtually impossible to put a full cord of wood on a pickup truck, and you are paying the price of a cord for what in reality is little more than 1/2 or 3/4 of a cord at best.

If you have a large amount of land in trees, you can cut only the trees that are dead or dying. You are simply pruning your treesI and eliminating the undesirable growth.

It is not a good plan, though, to cut indiscriminantly or to cut a tree simply because it is close to the road and therefore easier to haul to the truck. A good oak takes well over a century to develop into a mature tree. When a tree falls, it will not last long, so you should plan to cut it quickly, split it, and let it begin curing.

If you don't have a stand of trees, you may want to set out seedlings or make some arrangement with a neighbor who owns trees. You might agree to cut and haul away trees that are in the neighbor's way. If, as you are driving through your area, you notice trees that are dead and present a hazard or which have been damaged and should be cut, you might even find people who will pay you to cut the trees and haul away the wood. You could offer, if all else fails, to cut the tree in exchange for 1/2 of the wood.

As far as setting out seedlings is concerned, there is a type of poplar tree -- or tulip tree -- that is ready to cut and use within 5 years of the time it is set out. You can have your own stand of trees if you have the land area to spare.

Setting out trees is, in all estimations, a good investment, and you can improve the values of your property by adding the trees. Best of all, you add another link to your chain of self-reliance, and you have the comfort of realizing that, barring a calamity, you have a fuel supply assured.

Glossary

Glossary

air drying— a method of drying food in order to preserve it, in which food is exposed to open air and sunlight so that moisture can be evaporated. Also called solar drying.
antibiotics— any of certain chemical substances used in food preservation to inhibit the growth or to destroy bacteria and other micro-organisms.
ascorbic acid— Vitamin C.

bacteria— typically one-celled micro-organisms; some bacteria cause diseases and others are necessary for fermentation, nitrogen fixation, etc.
baking— cooking by dry heat, usually in an oven. When the term is applied to meats, it means the same as roasting.
barbecue— to roast meat in a slow fashion over hot coals and often on a spit or in an oven. Barbecued meats are usually basted with a seasoned sauce, although some are roasted in their own juices.
baste— to moisten foods during the cooking processes by ladling pan drippings, sauces, or water. Basting is used to prevent the foods from being excessively dry or to add flavor.
blanch— to pour boiling water over food, usually vegetables and fruits, to loosen peelings or to remove or set colors.
blancher— a large vessel containing boiling water. Foods to be preserved are lowered into the water.
blanching, steam— the process in which food is blanched by the steam of the boiling water instead of by pouring the water over the food and allowing it to set.
blend— to combine two or more ingredients or foods by mixing until smooth.
boil— to cook in a liquid at boiling temperature.
bottling— to store or preserve under pressure in a bottle, jar, or can. Also called canning.
botulism— poison resulting from the toxin produced by certain bacteria sometimes found in improperly preserved foods.
braise— to brown in a small amount of hot fat and then add a small amount of liquid and cook slowly in a tightly covered container on top of the range or in the oven.
bread— to coat or cover with bread crumbs or a mixture of bread crumbs and beaten eggs and milk, usually with the crumbs added as a final step.
brine— a heavily saturated salt solution, usually for use in pickling.
broil— to cook by direct heat, usually in an oven with a temperature setting over 500 degrees or over hot coals or flames.
Brix or Brix scale— a measurement of the density or concentration of sugar in a solution, usually with reference to the potential alcohol content of a wine fermentation.

campden tablet— a commercial preparation containing potassium metabisulphite, used as a sterilizing agent for wines. It inhibits the growth of wild yeast cells and other spoilage organisms that ruin wines.
candy— to cook in a sugar or syrup or to cook in a heavy syrup until fruit is plump and transparent.
canning— preserving food in airtight jars, bottles, or cans.
caramelize— to melt sugar slowly over low heat until it turns a dark brownish color.

carboy— a large glass bottle or jar, often holding as much as 5 gallons, used often for wine fermentation containers.
casein— a phosphorprotein that is one of the chief constituents of milk and the basis of cheese.
chaff— husks of wheat or other grains that are separated and discarded during threshing or winnowing.
charcoal— carbonized wood produced by burning wood partially and then extinguishing it by suffocation.
chitterlings— the small intestines of pigs or hogs, used for food. Often called 'chitlins.'
cholesterol— a fatty alcohol found in animal fats, blood, nerve tissue, and bile and thought to be a contributing agent in heart attacks.
chutney— a relish made of fruits, spices, and sometimes herbs.
cold-pack— a process of canning foods in which the raw products are first placed in jars and then subjected to heat.
conserve— a kind of jam made from two or more fruits, often with nuts or raisins added.
cord of wood— a measure of wood cut for fuel or pulpwood: 128 cubic feet as arranged in a pile 8 feet long, 4 feet high, and 4 feet wide.
corn shock— a bundle or circle of corn stalks stood erect, tied together, and left in a field to dry.
creosote— an oily liquid with a pungent odor derived from wood tar or coal tar.
cross-pollination— the transfer of pollen from the anther of one plant to the stigma of another, often through the action of wind or insects.
curd— the coagulated part of milk from which cheese is made; it is found when milk sours.
cure— to preserve meat or fish by salting or smoking usually.

decompose— to break up or separate into basic components.
dehydrate— to remove water content.
dice— to cut into small cubes.
dissolve— to cause a dry substance to pass into a liquid.
dredge— to sprinkle with flour or other fine substance.
drying— removing moisture from, through sun, air, or heat.

enzyme— a protein-like substance, found in plant and animal cells, that acts as an organic catalyst in initiating or accelerating a specific chemical reaction and that usually becomes inactive or unstable at high temperatures.

fermentation— a major step in wine-making or pickling processes in which there is a breakdown of complex molecules into organic compounds, caused by the presence of yeast, bacteria, or enzymes.
fermentation lock— a device used in wine-making to allow gas to escape from the wine and to prevent air from entering.
filet— a strip of meat, usually boneless; also boneless chicken or fish.
freezing— preserving foods by rapid refrigeration, reducing the temperature of the foods so that organic changes are impeded or halted.
fry— to cook in a pan, usually in grease or other similar substance, such as margarine or oil.

garnish— an additive to a plate of food to increase color or flavor.
glaze— to cover foods with a glossy coat of sugar syrup or gelatin.
graft— a shoot, bud, etc., of one plant artificially inserted into another tree or other growth.
grate— n. a metal construction to hold wood coals, etc.
grate— v. to chop or cut finely
gumbo— a soup thickened with okra.

hermetical sealing— sealing of cans or jars of foods so that no air or gas can enter or leave.
high acid— a term used to refer to the plentiful presence of acids in certain foods, such as tomatoes.
hornworms— the larval caterpillar of various hawk moths, with a horny growth on the final segment; a pest particularly damaging to such garden crops as tomatoes.
hybridization— the production of offspring by crossing 2 varieties of unlike genetic constitution.
hydrometer— an instrument for measuring the specific gravity of liquids; it is a graduated, weighted tube that sinks into a liquid up to the point determined by the density of the liquid.

Japanese beetle— an insect especially damaging to orchards, gardens, vineyards, and flowers and shrubs and resembling a miniature June bug.
jerky— dried beef.

kernels— the individual grains of corn on an ear or the edible part of several natural foods.

larding— the inserting, artificially, of fats into cuts of meat, primarily by use of a special needle.
low acid foods— neutral foods that, because of their low acid content, must be pressure canned, dried, or frozen.
lye— a strong alkaline solution obtained by leaching wood ashes or potassium hydroxide, an alkaline used in cleaning or soap-making.

marinate— to allow a food to stand in a liquid, usually a dressing, wine, or mixture of oil and vinegar, to add to the flavor or to tenderize.
Mason jar— a glass canning jar used for preserving foods, especially in home canning.
micro-organism— any of the bacteria, protozoans, viruses, etc.; a vegetable organism or ultramicroscopic animal.
milky spore spray— a treatment for the prevention of Japanese beetles.
moisture reduction— the loss of water or other liquid, particularly with reference to dehydration or drying of foods.
molds— a furry growth on the surface of organic matter, caused by fungi, especially in the presence of dampness or decay.

naphtha— a volatile oily liquid produced by fractional distillation of petroleum.
nectar— the sweetish liquid in many flowers, particularly attractive to bees and other insects.
nodule— a small knot or joint on a stem or root.
non-acid— see low acid.

open kettle— a food preservation process in which the food is cooked before canning and while hot is packed in hot jars and sealed.

oven drying— dehydration of food in the oven of the kitchen range, set on lowest heat, or in special ovens built for the purpose.

oxidation— the union of a substance with oxygen.

pan broil— to cook in a hot frying pan, usually greased, and pouring off fat as it is secreted.

pan fry— to cook in an open pan and usually in a small amount of fat.

paraffin— a waxy substance used for making candles and for sealing jars of food.

parboil— to boil in water until partially cooked.

pare— to peel or cut away the outer covering.

pasteurize— to preserve food by heating sufficiently to destroy bacteria, usually applied to milk and fruit juices.

pay-to-pick— a method of buying in which the purchaser agrees to select and pick his own fruits or vegetables.

pesticide— any chemical used for killing insects or weeds.

piccalilli— a relish of chopped vegetables, mustard, vinegar, and hot spices.

pit— to remove the hard stone from a fruit.

poach— to cook in hot liquid so that food holds its shape.

polyunsaturates— certain vegetable and animal fats and oils containing molecules with multiple bonds.

pressure canning— a method of canning in which jars are sealed by steam under pressure. Considered the safest process for canning non-acid vegetables, meats, fish, and poultry.

primary fermentation— first stage of fermentation in which yeast growth takes place and air reaches the must.

rehydrate— to restore water or other liquid to something that has been dehydrated.

relish— any of a variety of foods, especially pickles, olives, piccalilli, and raw vegetables, served with the main course of a meal or as an appetizer.

rennet— the membrane lining the fourth stomach of a calf.

rennin— a coagulating enzyme, found in rennet, that curdles milk.

roast— to cook by dry heat, usually in an oven.

root cellar— an underground storage room for vegetables, especially root crops.

root crop— any of a variety of crops, as turnips, beets, etc., grown for the edible roots.

root stock— a plant onto which another is grafted as a new top.

salt cure— to preserve meat, fish, etc. by heavily salting.

saute— to cook briefly in a small amount of hot fat.

scald— to heat to a temperature just below the boiling point.

scallop— to bake a food with sauce or other liquid.

scion— a shoot or bud of a plant that is suitable for planting or grafting.

score— to cut gashes in the skin or covering of food.

scramble— to stir eggs or an egg mixture while cooking until mixture sets.

sear— to brown a food, usually meat, very quickly over intense heat.

self-pollination— the transfer of pollen from anthers to stigmas in the same flower.

shock— the sheaves, as of corn or wheat, stacked together on end to cure and dry.

shread— to cut or tear in small, narrow pieces.

sift— to put dry ingredients through a sieve.

simmer— to cook slowly over very low heat.

smoke— to treat meat, fish, etc. with smoke in order to flavor or cure.

solar drying— process of drying food in open air by sunlight.

steam— to cook in steam of boiling water.

steep— to extract color, flavor, or other qualities from food by leaving it in water just below boiling temperature.

sterilize— to free from micro-organisms by subjecting to great heat, as in boiling, dry heat, or steam.

stew— to simmer in a liquid for a long period of time.

stock— the liquid in which meat or fish has been boiled, used as a base for soups and gravies.

sucker tomato— a subordinate shoot springing from the stem of a tomato plant.

sugar cure— to preserve meat, fish, etc. by heavily sugaring.

sulfur dioxide— a heavy, colorless gas used as a preservative.

sulfuring— treating with sulfur fumes in order to preserve, usually meats.

super— section of beehive containing frames of wax on which honeycombs are built.

tartar— dead yeast that appears as a thin substance on the bottom of the fermentation tank during the wine-making process.

tendril— a threadlike part of a climbing plant serving to support it by coiling around an object.

toast— to brown by direct heat.

tofu— a soybean product in curd form which can be fried, baked, poached, blended, marinated, chopped, stir-fried, or frozen, usually mixed with other foods for flavor enhancement.

toss— to mix ingredients lightly.

truss— to tie fowl or other meat with skewers to retain the shape of the meat during cooking

tuber— a thick, flesh part of an underground stem, as a potato.

vacuum seal— a seal obtained by steam under pressure during which most of the air was exhausted from the container being sealed.

Vidalia onions— extremely sweet onions grown only in Vidalia, Georgia.

viscera— the intestines.

water bath— a process of canning in which packed jars of food are lowered into a large vessel containing enough boiling water to cover jars completely during processing time.

whey— the thin, watery part of milk that separates from the curds, as in cheesemaking.

yeast— a microscopic fungus which consumes sugar, producing alcohol and carbon dioxide.

yogurt— a semisolid milk product which has been fermented with bacteria, believed to benefit the intestines, usually combined with flavors.

Index

Edited by Suzanne L. Cheatle